Bodies in Doubt

Bodies in Doubt
An American History of Intersex

ELIZABETH REIS

The Johns Hopkins University Press

Baltimore

For Matt, Sam, and Leah

© 2009 The Johns Hopkins University Press
All rights reserved. Published 2009
Printed in the United States of America on acid-free paper
9 8 7 6 5 4 3 2 1

The Johns Hopkins University Press
2715 North Charles Street
Baltimore, Maryland 21218-4363
www.press.jhu.edu

Library of Congress Cataloging-in-Publication Data
Reis, Elizabeth, 1958–
   Bodies in doubt : an American history of intersex / Elizabeth Reis.
      p. ; cm.
   Includes bibliographical references and index.
   ISBN-13: 978-0-8018-9155-7 (hardcover : alk. paper)
   ISBN-10: 0-8018-9155-8 (hardcover : alk. paper)
      1. Intersexuality—United States—History. 2. Gender identity—United States—History.
3. Intersex people—Identity—Social aspects—United States. I. Title.
   [DNLM: 1. Hermaphroditism—history—United States. 2. Gender Identity—United
States. 3. Health Policy—history—United States. 4. History, Modern 1601—United
States. 5. Social Conditions—history—United States. 6. Transsexualism—history—
United States. WJ 11 AA1 R375b 2009]
   RC883.R45 2009
   362.196'69400973—dc22   2008028739

A catalog record for this book is available from the British Library.

Special discounts are available for bulk purchases of this book. For more information,
please contact Special Sales at 410-516-6936 or specialsales@press.jhu.edu.

The Johns Hopkins University Press uses environmentally friendly book materials, including recycled text paper that is composed of at least 30 percent post-consumer waste,
whenever possible. All of our book papers are acid-free, and our jackets and covers are
printed on paper with recycled content.

CONTENTS

ACKNOWLEDGMENTS

I have received enthusiastic support for this project from friends, colleagues, and family.

I am indebted to the readers at the *Journal of American History* and David Nord for their comments on "Impossible Hermaphrodites: Intersex in America, 1620–1960." I also appreciate the almost instantaneous and enormously helpful evaluations I received for the article that I published in *Perspectives in Biology and Medicine*, "Divergence or Disorder? The Politics of Naming Intersex," which is the epilogue of this book. Special thanks go to David Iris Cameron, Milton Diamond, Alice Dreger, Katrina Karkazis, Emi Koyama, Bo Laurent, Iain Morland, Bob Perlman, David Sandberg, Paul Vasey, and Eric Vilain. Among these, Alice Dreger must receive special mention. I am deeply thankful for her thorough, perceptive reading of the entire book and her unreserved encouragement.

Others have talked with me about the project or read portions of the manuscript and offered their suggestions and critique. I would like to thank the Pacific Northwest Early American Reading Group, the students in my Sex and Medical Ethics class at the University of Oregon, Lee Arbogast, Susan Armeny, Linda Barnes, Diane Baxter, Andy Burstein, Cynthia Eller, Tom Foster, Lynn Fujiwara, Richard Godbeer, Nancy Isenberg, Lauren Kessler, Suzanne Kessler, Catherine Kudlick, Susan Lantz, Ernesto Martinez, Christina Matta, Joanne Meyerowitz, Julie Novkov, Lynn Nyhart, Catherine Johnson-Roehr, Susan Johnson-Roehr, Robert Nye, Judith Raiskin, Ellen Scott, Do Mi Stauber, Susan Stryker, Kate Sullivan, Jennifer Terry, Kerry Wilson, Mac Wilson, and Al Young. I would also like to thank audiences and panelists at the Kinsey Institute Conference, the American Association for the History of Medicine, the Western Association of Women Historians, and the Organization of American Historians Meeting as well as listeners at Cornell University, University of Miami,

University of Oregon, Oregon State University, Southern Oregon University, and Mother Kali's Bookstore in Eugene, Oregon.

I have received financial support from the University of Oregon Humanities Center and the Center for the Study of Women in Society. Arlene Shaner of the New York Academy of Medicine helped a great deal while I was researching the book and subsequently as I tracked down sources or needed permissions to reprint pictures. Thanks also to Sara Brownmiller at the University of Oregon, Shawn C. Wilson and Catherine Johnson-Roehr at the Kinsey Institute for Research in Sex, Gender, and Reproduction, Brenda Marson at Cornell University, and Florence Gillich at Yale University's Medical Historical Library.

My mother, Pamela Tamarkin Reis, deserves a prize. She meticulously read and refined every chapter more than once, even though the topic is far from her own field of study. My children, Sam and Leah Reis-Dennis, were troopers. Living in a house where books having to do with bodies and sex proliferate can be challenging for teenagers. Once my daughter did a Google image search for my name in the school library and was taken aback when pictures of ambiguous genitalia from an article I had published appeared as well. As teens though, they can certainly appreciate issues of autonomy and consent, and I hope that my work on this project has benefited them in some way. My husband, Matthew Dennis, kept me thinking like a historian, always prodding me to set intersex issues in a broader context and to consider the bigger picture of American history. And of course, Matt, Sam, and Leah often lured me away from my desk to take breaks, play, and focus on the present, for which I am enormously grateful.

INTRODUCTION

What does it mean to be human? Since antiquity, philosophers have pondered such a question. We need not answer the question comprehensively, nor do we need to be philosophers to realize that among the essential attributes of humanity is sex. To be human is to be physically sexed and culturally gendered. Indeed, historically societies have divided humans into males and females based on the nature of their bodies and structured social life on such a basis. In the United States and most other places, humans are men or they are women; they may not be neither or both. Yet not all bodies are clearly male or female. My book examines this anomaly—not the "anomaly" of bodies that do not easily fit into the categories of male and female (a more common condition than many suppose) but, rather, the anomaly of social relations in colonial America and the United States that could not manage greater flexibility and acceptance of bodily diversity among its members. From the beginning, intersex has been understood to be a problem. *Bodies in Doubt: An American History of Intersex* probes this "problem" and analyzes how its nature has changed from early America to the present. How do we—those with intersex bodies, the public, medical professionals—regard and treat intersex people today? How did they do so in the past? Most importantly, how has such regard and treatment changed over time, and how do those shifting understandings and approaches inform our sense of the gendered boundaries of American life and humanity itself?

Given the efflorescence of scholarship on gender and sex during the last generation, it is surprising perhaps that no one has attempted to address the history of intersex in America in any comprehensive fashion, particularly because intersex stands at the intersection of sex and gender and has been critical in defining the exclusive boundaries between male and female.[1] This volume examines the changing definitions, perceptions, and medical management of "hermaphrodites" (a term used historically) from the colonial period to the present.[2] Rather than focusing on the development of medical protocols or technical improvements in surgical procedures, I analyze the cultural history of American doctors and laypeople as they considered bodies and identities that fell outside the conventional categories of male and female.

Medical practice cannot be understood apart from the broader culture in which it is embedded. As the history of responses to intersex bodies has shown, doctors have been and continue to be influenced by the values and anxieties of the larger society, which render any medical management a cultural, rather than simply a scientific, endeavor.[3] Some nineteenth-century doctors, for example, approached their intersex patients with disrespect and suspicion, regarding them as willfully deceptive and insincere. Later doctors tended to show empathy for their patients' plight and displayed a desire to make correct decisions regarding their care. Yet definitions of "correct" in matters of intersex were entangled with shifting ideas and tensions about what was natural and normal, indeed, about what constituted personhood or humanity. Patients were just as ensnared in this world as doctors. Though medical sources give us only a selective glimpse of patients' ideas and wishes, we can occasionally see how intersex people too reflected and expressed the social norms of their time and place.

Throughout the period under study, many doctors and laypeople believed that hermaphroditism, as it was then defined, did not exist in the human species. No human being had ever been found with perfect sets of both male and female sexual organs. Technically, the doctors were right: no humans mirrored, say, hermaphroditic earthworms, possessing two perfect sets of external and internal reproductive organs, capable of reproducing as either female or male. Hermaphroditus, the figure from Greek mythology whose male body was merged by the gods with the female body of the nymph Salmacis, found no counterparts in the human world. But saying that hermaphrodites did not exist encouraged doctors and laypeople to insist on two and only two sexes, when not all bodies fit precisely into discrete male and female categories. True, most people have bodies with physical markers that are clearly male or female, but some are born with genitals, gonads, and genetic material sufficiently equivocal to make doctors and parents wonder to which sex they belong.

Intersex generally refers to variation in genital anatomy, but not all intersex conditions involve ambiguous genitalia. Some people with intersex have typical external genitals but the internal anatomy of the other sex. Sometimes they do not find out about their condition until the teenage years, when their bodies do not go through puberty in the usual manner. There are several dozen congenital conditions, including (but not limited to) hypospadias, Turner syndrome, congenital adrenal hyperplasia, androgen insensitivity syndrome, 5alpha-reductase deficiency, and sex chromosome mosaicism, that fall under the rubric of intersex.[4]

Medical experts estimate that one of every two thousand people is born with genital anomalies. Our gendered world forces us to put all people into one of two categories when, in fact, as the biologist Anne Fausto-Sterling has suggested, we need to consider the less frequent middle spaces as natural, although "statistically unusual."[5] As we shall see, doctors have spent much time trying to determine intersexed people's "true sex" and molding their bodies to reflect their judgments, which have been based on social factors as much as (or more than) physical ones.

Chapter 1 describes early American responses to hermaphroditic bodies and explores colonial medical and legal understanding of hermaphrodites, especially as it intersected with concerns about "monstrous" births, marriage, impotence, privacy, and same-sex sexuality. Ministers and medical practitioners saw providence or the diabolical in "monstrous births." Mothers were often blamed for their babies' anatomies, for folk wisdom taught that maternal imagination could cause birth anomalies. Medical men and midwives made pronouncements about gender that had extensive legal ramifications—whether to settle cases of bastardy or impotence, confirm or deny rape accusations, or substantiate divorce proceedings.

Medical jurisprudence manuals and gynecological treatises of the eighteenth century doubted the existence of "perfect" hermaphrodites and instead insisted that many so labeled were simply women with enlarged clitorises. Such mistakes arose, some medical authorities suggested, from ignorance of human anatomy, particularly of female anatomy. Many who diagnosed a large clitoris believed that this condition presaged a sexually dangerous situation that could lead to two evils; it could hinder coitus and promote sexual relations between women.

Chapter 2 shows how the newly authoritative and professionalized physicians of the nineteenth century took the liberty of assigning a sex to people with uncertain genitals, even if that assignment opposed the patient's preference and contradicted the individual's previous gender performance. Like the rest of nineteenth-century society, doctors held the institution of marriage in high esteem and attempted to construct genital organs that might accommodate normative marital relations. There were also legal reasons for "proving" patients either male or female, whether the question was one of enfranchisement, inheritance, or divorce. Chromosomes and hormones were unknown, and internal inspection of gonads was not yet possible. Confounded physicians could but observe the often contradictory genitalia, and consequently their pronouncements depended as much on traditional stereotypical social indicators as on biological markers when they decreed a physically anomalous individual male or female. Nevertheless, decree they did, for convention required that every person be clearly male or female.

Cultural concerns about race also influenced physicians' opinions. In the nineteenth century many of their published articles about intersex conditions describe African Americans with various malformations, as if to suggest that monstrosity (an idea that lingered from an earlier era) and blackness went hand in hand. In addition, the disquieting prospect that individuals could suddenly change sex, as some hermaphrodites seemed to do, paralleled the early national preoccupation with racial classification and the possibility (and fear) of mutable racial identity. Chapter 2 discusses this cultural preoccupation with the instability of identity. Whether it was to ensure the legal status of men or women, or to present sex, like race, as something uncomplicated, permanent, and easy to determine, in this era (before the ubiquity of genital surgery) nineteenth-century doctors insisted on certainty rather than ambiguity in gender designation.

In the late nineteenth and early twentieth centuries, the period discussed in chapter 3, the new paradigm of sexual inversion affected the medical interpretation of hermaphrodites. The idea of sexual inversion had emerged as a scientific explanation of homosexuality. By the turn of the century, hermaphrodites were considered potential homosexuals or "inverts"; if some people's bodies could look both male and female, then would such individuals be attracted to the "wrong" sex? Though the possibility of hermaphrodites being physically intimate with persons of either sex had long concerned physicians, American doctors now began to evaluate their patients' sexual inclinations and to intervene to surgically ensure that sexual intercourse, when it occurred, would take place between two differently sexed bodies. Doctors wanted genitalia to match heterosexual desire. If a patient with ambiguous genitals expressed a sexual interest in women, surgeons would try to ensure that surgically "repaired" male genitals could penetrate. Similarly, if the patient showed sexual interest in men (or expressed no sexual desire, for doctors often considered the sexual urge to be a male, not a female, impulse), fashioning female genitalia became the project. Such privileging of heterosexuality persisted throughout the twentieth century among physicians and laypeople alike, and current intersex activists have critiqued its impact on intersex people.

Chapter 4 relates the difficulties caused by the medical establishment's eventual embrace of the gonadal standard of sex determination, and it chronicles the rising importance of psychology in deciding a person's sex. During the 1920s and 1930s, doctors battled with each other in print over the defining markers of sex. If gonads, discovered surgically, were not the most important indicator, then what was? In the first three decades of the twentieth century, ethical questions multiplied, as doctors who believed that the gonads spoke the final word on biological sex confronted patients whose gonads contradicted their genitals. What should physicians do if they saw an obviously, or predominately, female patient who had internal testes? Did the presence of male gonads make one "really" a man? Should they tell someone who had always lived as a woman that she was a man? As in earlier periods, a commitment to heterosexual marriage biased many of their decisions. If, for example, a woman with internal male gonads wanted to marry a man, then her doctors would be likely to promote her femininity, so that she could become heterosexually married. Doctors' primary motivation was not to enable the patient to have the partner s/he wanted, but rather to promote broader cultural values such as marriage.

In the 1940s and 1950s acceptance of psychology as a scientific endeavor grew, and psychological testing became part of sex determination for adult patients. The emphasis on the gonads declined as the reliance on psychology increased. Doctors focused their attention on how best to ensure mentally healthy adult patients (part of the definition of "mentally healthy" remained heterosexual). The psychologist John Money of the Johns Hopkins University had written his dissertation on the psychological health of people with ambiguous genitalia. He discovered that people born with intersex conditions were surprisingly psychologically vigorous, a finding that seemed to challenge common sense. Here were people who had grown up with supposedly troubling incongruities between their anatomies and their gender identities, and yet they seemed to do so with their psyches remarkably intact. Money surmised that the most successful patients would be those who had been raised as unequivocal boys or girls, despite their contradictory genitals.[6] Diminishing genital incongruity in infancy so that parents could rear unambiguously identified children became the goal of intersex management for the next fifty years.

Chapter 5 takes the story to the late twentieth century, when doctors across the country routinely treated neonatal cases of ambiguous genitals as medical emergencies, often abandoning previous standards and almost invariably performing surgery to "correct" external imprecision, whatever the gonadal evidence. The determination to make intersex bodies look like those of typical males or females reached its apogee through the influence of John Money. He held that a person's sense of gender identity was malleable until about eighteen months of age. He therefore concluded that those born with ambiguous genitalia could have their sex surgically assigned as infants and later hormonally regulated without negative consequences. Though it had some critics, Money's theory enjoyed almost unprecedented acceptance and adherence for decades.

Through the mid-twentieth century, the history of intersex in America has been a largely untold story. My work on the latter half of the century joins a growing body of scholarship fueled, in part, by critiques from intersex people themselves. In the last fifteen years, intersex activists, former surgical patients (sometimes the same people), and others have come forward to say that Money and his colleagues were wrong. Some children thus altered never felt at home in their assigned gender.[7] Some of them did not know their medical histories, as doctors had advised parents and relatives to keep the matter secret. When they found out, a number were relieved and changed gender as adults. Many stayed with the gender chosen at birth but struggled to accommodate to life with surgically altered sexual organs that had been severely compromised, burdened by a deep sense of shame induced by enforced secrecy. As patients grew older and endured further surgical and hormonal interference, doctors' objectives remained consistent with historical aspirations: eliminate monstrosity, shape normality, and ensure patients' marital prospects. Today, Money's findings have been discredited, and the injunction to "wait until puberty and see" that is gaining acceptance surely obviates much heartache. Physicians believe that their current advice to choose a gender but avoid surgery depends not on ideology or fear of eventual legal action but on empiricism. The primary obligation in medicine is, first, to do no harm, and this new protocol, unlike those of previous centuries, seems to obey that requirement.

The epilogue examines the current (and perhaps ongoing) naming controversy. The term "hermaphrodite" continues to linger but is denigrating. "Intersex" is more neutral and positive, but some parents object that it marks their child as "other," a third sex, something in between male and female, while they will be raising their child as either a boy or a girl. In addition, the term suggests sexuality, and many new parents are uncomfortable thinking about their baby's future sex life. The new medical term "DSD" (disorders of sex development), takes some of the attention away from matters of sexuality, gender, and monstrosity, particularly because of the availability of the acronym, but I and others think that it unnecessarily medicalizes a situation that might not need medical care. The vast majority of intersex conditions are not life threatening. Not everything has to be "fixed," especially when the fix is based on social concerns rather than medical necessity. In this epilogue I suggest a new term, still DSD, but standing for divergence of sex development. By using this nomenclature, doctors, patient, and parents can articulate difference but not disorder.

How unusual bodies are treated is a critical historical question. In the United States they have been marked as "other," as monstrous, sinister, threatening, inferior, and unfortunate. Once it was respectable to call those with unusual physical forms monsters, later not. But the mingling of fascination and fear lingers, and the refusal to be "normal," whatever that is, triggers hostile expressions. We might imagine other possibilities, perhaps drawing on path-breaking work in the new history of disability that invites scholars to ask how atypical or confusing bodies became "abnormal" bodies. My book therefore considers how Americans have understood and handled ambiguous bodies, how the criteria and the authority for judging such bodies changed, how both the binary gender ideal and the anxiety over uncertainty persisted, and how the process for defining the very norms of sex and gender evolved.

A NOTE ABOUT THE ILLUSTRATIONS

Some contemporary intersex people oppose medical photography because of its tendency to expose and objectify, reducing subjects to body parts only. Intersex people who have been photographed unremittingly at doctors' visits have at times felt exploited and have commented on the sense of shame that such scrutiny engenders.* Sensitive to these concerns, I struggled with the questions of whether to include historical photographs. Despite some reservations, I have decided to include them. I excluded pictures taken after the 1940s, based on the likelihood that adult subjects photographed earlier in the twentieth century would no longer be alive. Additionally, I have included only photographs that I believe serve a scholarly purpose—sometimes to illustrate how doctors came to their conclusions regarding a person's gender, other times to reveal how doctors used pictures to justify their medical treatments and decisions. As with the textual documents, I analyze the photographs critically. I sincerely hope they do not offend readers; they are meant to illuminate doctors' past motivations, not to endorse them.

*S. Creighton, J. Alderson, S. Brown, and C. L. Minto, "Medical Photography: Ethics, Consent, and the Intersex Patient," *BJU International* 89 (2002): 67–72.

Bodies in Doubt

CHAPTER ONE

Hermaphrodites, Monstrous Births, and Same-Sex Intimacy in Early America

I departed [Newcastle, Delaware] thence at half an hour after three, and about a mile from town I met a monstrous appearance, by much the greatest wonder and prodigy I had seen in my travels, and every whit as strange a sight by land as a mermaid is at sea. It was a carter driving his cart along the road, who seemed to be half man, half woman. All above from the crown of his head to the girdle seemed quite masculine, the creature having a great hideous unshorn black beard and strong coarse features, a slouch hat, cloth jacket, and great brawny fists, but below the girdle there was nothing to be seen but petticoats, a white apron, and the exact shape of a woman with relation to broad round buttocks. I would have given something to have seen this creature turned topsy-turvy, to have known whether or not it was an hermaphrodite, having often heard of such animals, but never having seen any to my knowledge; but I thought it most prudent to pass by peaceably, asking no questions, lest it should prove the devil in disguise.

So WROTE DR. ALEXANDER HAMILTON in the mid-eighteenth century. His description of a possible hermaphrodite sighting is meant to be amusing and fantastical. A hermaphrodite? The very idea was preposterous, for such a creature was as mythical as a mermaid, he implied. But Hamilton's far-fetched vignette hints at tropes found in more serious appraisals of hermaphrodites that I will discuss in this chapter.[1] The early modern definition of a hermaphrodite (one person with perfect sets of both male and female genital organs) effectively precluded anyone from fitting the category. That led some medical authorities to believe that hermaphrodites did not exist in the human species. Nevertheless, some people were born with ambiguous genitalia or with body parts that allowed them flexibility in living variously as male and as female. Though their bodies might not fit the criteria exactly, such people were sometimes considered hermaphrodites. While Dr. Hamilton expressed incredulity, wonder, and curiosity at the idea of a fabled creature, others in early America seriously pondered the possibility of hermaphrodites' existence and set out rules for interaction with those who approximated the standard.

Before "hermaphroditism" became understood as a medical condition that inevitably necessitated medical intervention, early medical, legal, and religious authorities had opinions on how people with atypical genitalia should be regarded. In this chapter I explore responses to hermaphrodites, ranging from (as Hamilton suggested) ideas that such "monstrous" births signaled their parents' sinful nature to worries that hermaphrodites were really women with enlarged clitorises, potentially able to copulate with other women. Even if, according to medical authorities, hermaphrodites did not exist, the idea of *one* body exhibiting *two* sexes, able to couple with either sex, raised a host of anxieties about gender, sex, and sexuality. Throughout the early period we confront the issue of choice; a binary system of sex, the ideal established and authorized by the biblical Adam and Eve, was rigid, and choosing only one for each individual (despite ambiguity and contradictory markers) was mandatory.

In early America, doctors did not have the social status or the medical knowledge that they acquired in the nineteenth century, and common people typically managed illness and disease without professional help.[2] Because most people tended to their own health needs, extensive medical records are not available to provide historical sources on intersex in the era. Later, in the early nineteenth century, as doctors professionalized, they wrote journal articles about conditions their patients endured, including atypical genital anatomies, and since then historians have had rich medical material to interpret.

Some American midwives and doctors read European medical manuals, and so their understanding of diverse conditions was no doubt influenced by a European intellectual tradition going back centuries. The handful of early American authors who wrote their own books cited the

European writers, whether or not they agreed with their ideas. Early Americans with no pretensions to expertise also read European and British treatises; some of these books, like Jane Sharp's *The Midwives Book; or, The Whole Art of Midwifery Discovered* and Nicholas Culpeper's *The Compleat Practice of Physick* became popular in the colonies. Fortunately, medical treatises are not the only available sources. Sermonic literature reflecting the religious interpretation of illness and disability exists, at least from New England, and throughout the colonies there were legal records, which sometimes involved charges of impotence related to what we now see as intersex conditions. In addition, newspapers and literary sources, such as Hamilton's *Itinerarium*, can offer clues as to how colonists understood hermaphrodites in an era before hermaphroditism was considered a pathology requiring treatment.

Monstrous Births

Early Americans placed hermaphrodites in the broad category of monstrous births, a catchall designation that included all kinds of birth anomalies.[3] Monsters, few doubted, were sent by God as signals and warnings. Puritan theologians agreed on the doctrine of providence; since God ordered the universe, every unusual event had divine significance.[4] Sometimes God sent subtle signs; other times, depending presumably on the importance of the occasion, his messages were more obvious. Mary Dyer's unfortunate malformed baby was one such blatant expression, understood to be a dramatic expression of God's censure. Dyer had been a follower of Anne Hutchinson (a woman expelled from the Massachusetts Bay Colony for criticizing established clergy, expounding the doctrine of "grace in the heart," and testing the limits of female autonomy in matters of faith) and later became a Quaker. In 1637 she gave birth to a terribly misshapen baby that John Winthrop, the colony's governor, described like this:

> It was a woman child, stillborn, about two months before the just time, having life a few hours before; it came hiplings till she turned it; it was of ordinary bigness; it had a face, but no head, and the ears stood upon the shoulders and were like an ape's; it had no forehead, but over the eyes four horns, hard and sharp; two of them were above

one inch long, the other two shorter; the eyes standing out, and the mouth also; the nose hooked upward; all over the breast and back full of sharp pricks and scales, like a thornback; the navel and all the belly, with the distinction of sex, were where the back should be, and the back and hips before, where the belly should have been; behind, between the shoulders, it had two mouths, and in each of them a piece of red flesh sticking out; it had arms and legs as other children; but, instead of toes, It had on each foot three claws, like a young fowl, with sharp talons.[5]

The Reverend John Cotton suggested that Dyer conceal the birth. He saw "a providence of God in it . . . and had known other monstrous births, which had been concealed, and . . . he thought God might intend only the instruction of the parents and such other to whom it was known, etc." But God apparently wanted all to know, for during labor, about two hours prior to the birth, the bed shook violently, and Dyer's body emitted a "noisome savor"; the shaking and foul smell indicated that Satan lurked nearby.[6] Most of the women attending Dyer "were taken with extreme vomiting and purging" and were forced to leave.

Though Winthrop assigned blame to Dyer only tentatively (he never explicitly tied her unsuitable religious leanings to the stillborn baby), Puritan theologians saw a more direct correlation between monstrous births and dangerous opinions. In a sixteenth-century English text dealing with such births, the conclusion was unequivocal. These births, it explained, "signifie the monstrous and deformed myndes of the people mysshapened with phantastical opinions, dissolute lyvynge, licentious talke, and such other vicious behavoures which mounstrously deformed the myndes of men in the syght of god who by suche signes dooth certifie us in what similitude we appere before hym, and thereby gyveth us admonition to amende before the day of his wrath and vengeance."[7] Dyer no doubt participated in all four sins specified: fantastic opinions, dissolute living, licentious talk, and other miscellaneous vicious behaviors such as her addiction to revelations, or what was seen as supernatural communication. As a result, she received God's explicit physical condemnation in the figure of the deformed child. The minister Thomas Weld's account of Dyer's daughter's birth concurred: "Then God himselfe was pleased to step in with his casting voice . . . by testifying his displeasure against their [Dyer's and Hutchinson's] opinions and practices, as clearly as if he had pointed with his finger."[8]

Winthrop's horrific description of Mary Dyer's baby conformed to other colonial portrayals of birth anomalies. *Aristotle's Master-Piece,* the most popular and often reprinted medical manual in the colonies and the definitive word on all matters relating to reproduction and genital anatomy, included accounts of creatures born with composite features resembling animal, human, and mythic forms.[9] The book detailed a case in 1393, for example, when a woman copulated with a dog, producing a beast that resembled the mother from the waist upward and the dog, with paws and a tail, from the waist down. Next to its depiction of that being, the *Master-Piece* showed a similarly composite hermaphroditic creature, thus explicitly linking inhuman monsters and hermaphrodites for readers. The latter "monster," as the *Master-Piece* termed it, was born in 1512 in Ravenna, with a horn on the top of its head and two wings; it stood on only one foot, which had talons, like a large bird. It had "the Privities of Male and Female, the rest of the Body like a Man, as you may see by this Figure."[10] The creature is not specifically labeled a hermaphrodite, but the description of its "privities" conforms to the colonial understanding that a hermaphrodite had both male and female genitals.

There were gendered explanations for monstrous births of all sorts; irregularities of various kinds were often attributed to maternal imagination, for example. According to the *Master-Piece,* a pregnant woman's unruly thoughts could cause a birth anomaly. If a pregnant woman saw a rabbit, for instance, her child might be born with a harelip (cleft lip), the manual cautioned. "Some Children are born with flat Noses, wry Mouthes, great blubber Lips, and ill-shap'd Bodies; and most ascribe the reason to the Imagination of the Mother, who hath cast her Eyes and Mind upon some ill-shap'd Creature."[11] Pregnant women were cautioned to steer clear of such sights, or at least to avoid staring at them.[12]

Jane Sharp's *The Midwives Book; or, The Whole Art of Midwifery Discovered,* one of the few midwifery manuals written by a woman, had also cautioned against women's excessive imagination during pregnancy. "Sometimes the mother is frighted or conceives wonders, or longs strangely for things not to be had, and the child is markt accordingly by it," she wrote. Sharp told a widely circulated story of a white woman who had a dark-skinned child; the woman had "lookt on a Blackmore [and] brought forth a child like to a Blackmore." Like other authors, Sharp included observations from her own experience, a practice no doubt meant to persuade readers of their veracity. "One [woman] I knew," she recalled, "that seeing a boy with two thumbs on one hand, brought forth such

Many editions of Aristotle's *Master-Piece*, a famous colonial medical manual, included pictures of monstrous creatures reportedly born to women. Though not specifically called a hermaphrodite, the "Winged Monster" had "the Privi-ties of Male and Female." Image from Aristotle's *Master-Piece, Completed in Two Parts: The First Containing the Secrets of Generation in All the Parts Thereof* (London, 1700). Courtesy of the Kinsey Institute for Research in Sex, Gender, and Reproduction.

another." Such births could have other explanations, to be sure, but Sharp insisted that "the imagination is so strong in some persons with child, that they produce such real effects that can proceed from nothing else; as that woman who brought forth a child all hairy like a Camel, because she usually said prayers kneeling before the image of St. John Baptist who was clothed with camels hair."[13]

The hairy offspring and "Blackmore" stories (and accompanying woodcut) appeared in several medical guides, including *Aristotle's Master-Piece*.[14] According to the historian Mary Fissell, the hairy woman conjured notions of both lust and animality. Readers understood that women could copulate with beasts, resulting in monstrous births. Women could also simply imagine copulating with animals, or, as Sharp suggested, just think about an animal asexually, and the results could be equally disastrous. The connection between race and monstrosity was also not so subtly suggested. As Fissell has shown, including the small black child in various editions of the *Master-Piece*'s frontispiece encouraged readers to connect women's sexual imaginations, racialized fantasies, and racial mixing. The 1700 edition of the *Master-Piece* cautioned: "And I have heard of a Woman, who at the time of Conception, beholding the Picture of a Black moor, conceived and brought forth an Aethiopian."[15]

To avoid monstrous offspring, the *Master-Piece* also advised women to abstain from sex during their menstrual periods. "Undue copulation," the book explained, was "unclean and unnatural." The "issue of such Copulation does oftentimes prove Monstrous, as a just Punishment for their Lying together, when Nature bids they should forbear." Women and men were both to blame for bad coital timing, though most of the onus was put on women to prevent such activities; men were faulted for being too eager for intercourse, but women should know their bodily condition and, if necessary, refuse their partners.[16] Whoever was to blame, the *Master-Piece* made clear that one cause for monstrous births was divine: "Because the outward Deformity of the body, is often a Sign of the Pollution of the Heart, as a Curse laid upon the Child for the Parents Incontinency."[17]

In early American medical texts, authors usually discussed hermaphrodites in relation to monstrous births. The discussion of monstrosity typically included three questions: "What is the cause of monsters? Whether they are possessed of life? Whether a perfect monster can be considered a human being?"[18] *Aristotle's Master-Piece* maintained that those born by "natural means," as opposed to an "unnatural" union between a woman and a beast, "tho [sic] their outward Shape may be deformed and monstrous; have notwithstanding a reasonable Soul, and consequently their Bodies are capable of a Resurrection."[19] These same questions and more, both theological and medical, were asked about hermaphrodites. In answering them, most writers acknowledged the humanity of those born "imperfect monsters," in which only the genitals were affected.

By the mid-eighteenth century, some writers were turning away from the supernatural as either an explanation for unusual births or a way to reclaim a certain dignity for those with atypical bodies. Natural arguments based on Enlightenment assumptions replaced them. James Parsons, the British author of an eighteenth-century medical text devoted to proving that hermaphrodites did not exist in the human species, used the new vocabulary to lament the sorry fate that befell those deemed more monstrous than human. Here he chastised Americans specifically: "Innocent children have been punished, and even put to Death, for having been reputed Hermaphrodites," he mourned. "Ignorance of the Fabrick of the Body has been the first great occasion of those Evils, destroying Evils, which exist not only amongst most ignorant Americans, but also amongst the Litterati themselves in other Parts of the World." According to Parsons, even "Men of Science" believed reports of hermaphrodites, perhaps not unlike Alexander Hamilton's fanciful account. And he hoped that only the "weak-minded" would be persuaded by such tales, which he thought inspired people to "lose all Humanity towards such Objects."[20]

Ambiguous Sex, Impotency, and Privacy

Just as accounts of monstrous births offer a glimpse of colonial thinking about people with nonconforming bodies, so do prosaic legal records. Accounts of divorce proceedings for impotence, which combine both legal and medical interpretations of intercourse and marriage, are a fertile source for historians. Charges of impotence provoked physical examination of the impugned husbands. Some such colonial cases revealed what we might today consider intersex conditions, exposing not only the husband's failure to perform sexually but also his physical anomaly. In June 1686, for example, Dorathy Clarke of Plymouth, Massachusetts, petitioned the court for a divorce, stating that her husband, Nathaniel Clarke, "hath not performed the duty of a husband to me." Dorathy alleged that her husband was "misformed" and that he was "always unable to perform the act of generation." She requested a divorce because their "lives are very uncomfortable in the sight of God." Nathaniel denied the charges of "infirmity of body," and so the court ordered that "his body be viewed by some persons skilfull and judicious." The court chose three male physicians to inspect Nathaniel's body and give their judgment at the next court date. The findings of the physicians are not clear, but one month later the court decided that Dorathy would not be granted the divorce she requested.[21]

Impotence, particularly if attributed to a "misformed" penis, was regarded as a potential indicator of a hermaphroditic condition. Bodily examinations were common in such cases. The three doctors who scrutinized Nathaniel Clarke may have been looking for an unusually small penis that might have hindered sexual intercourse or a malformation known as *hypospadias,* where the urethral opening was on the underside rather than the tip of the penis. These conditions were recognized and, according to colonial law, would have been reason enough for a divorce, since they presumably predated the marriage contract. As the divorce was not granted, Nathaniel Clarke must have displayed some lesser (and acceptable) physical anomaly that left him capable of coitus or, alternatively, fully normal genitalia.

In 1662 a Massachusetts court heard a similar case, in which the husband admitted his impotence. Mary White sought a divorce from her husband, Elias White, because he "cannot performe the duty or office of a husband to hir." The court "perused the evidence" and did not see sufficient cause to separate the couple. Instead, the court advised them to work harder at their marriage. The husband appended a note to the court documents attesting to the truth of his wife's charges. He explained that when he first married he thought himself "sufficient: otherwise I neuer would have entered into that estate." Later he came to discover that he was "Infirmous not able to performe that office of marriage," though he could not determine the cause. Two men questioned Elias and Mary about the husband's sexual performance. When White lay with his wife, they asked, was "there any motion in him or no?" He answered that sometimes, after lying together four or five hours, there was, but "when he turned to hir It was gonn againe." Mary White asked her husband "whither or no he had euer made use of hir," and he answered "no."[22] Here, too, the court ruled against the divorce, perhaps because White agreed that when he married, he considered himself "sufficient." In other words, his infirmity became known only after the couple had been married for several years. As there was no fraud in the initial contract, a divorce on these grounds would not have been warranted.[23]

Women with congenital malformations of the genital organs could also be ruled impotent. And they too endured physical scrutiny by doctors and midwives to see if such conformation was causing their sexual problems or sterility. Early nineteenth-century American doctors combed the published records of European doctors, searching for cases that would help them diagnose the variety of genital malformation and its effects. American medical treatises include many examples of seventeenth- and eighteenth-century European women who sought medical (and often legal) attention because their marriages could not be consummated. Theodric Romeyn Beck and John Beck, authors of a leading medical jurisprudence textbook, described a Parisian woman who married in 1722 at age twenty-five but had not achieved intercourse for six years because, she said, "she could find none of the sexual organs, and that their place was occupied by a solid body." A surgeon was called in to evaluate, and he made an incision in the mass, which he thought would alleviate her condition, but to no avail. Twenty years later, in 1742, the husband sought to annul the marriage. As in the cases noted above, the woman's body was open to scrutiny. Two doctors found an "aperture of two or three inches" left from the previous surgeon's efforts, but they agreed that "either through fear or the prudence of the surgeon," the mass had not been entirely removed. Nonetheless, the court refused the annulment, on the grounds that the woman's situation was operable. Despite the failed earlier attempt, the court insisted that a "cure" was possible.[24]

The people in the aforementioned Massachusetts cases may have had intersex conditions that prevented sexual relations for one or both partners, but the term "hermaphrodite" was not raised or implied in court, perhaps because the litigants lived their lives uncomplicatedly as either men or women.[25] The first explicit case of ambiguous sex found in early American legal records is that of Thomas/Thomasine Hall, who was apprehended in Virginia and came before the court in 1629 for "dressing in women's apparel."[26] Hall's indefinite gender performance matched his or her bodily conformation, and various people took it upon themselves to inspect Hall's genitals and render a verdict as to whether Hall was a man or a woman. One man, in fact, cried out to Hall, "Thou hast beene reported to bee a woman and now thou art proved to bee a man, I will see what thou carriest." The deposition then describes the ensuing violation of Hall: "Whereuppon the said Rodes laid hands upon the said Hall, and this examiner did soe likewise, and they threw the said Hall on his backe, and then this examiner felt the said Hall and pulled out his members whereby it appeared that hee was a Perfect man."[27] Hall was commanded to "lye on his backe" and show his genitals many times during his ordeal. Hall was searched by both male and female investigators; one time two men even came into his room while he slept to sneak a look. Perhaps Hall resisted and the fact was not recorded, or perhaps he tolerated close scrutiny because such intimate inspection of people suspected of crime was not unusual. Searches of suspected "witches" occurred throughout the colonies as well, with examiners looking for the devil's mark or a teat whereby the devil's familiar could suck from the witch's body.[28]

Individuals born with ambiguous genitals, even if they were not pronounced perfect hermaphrodites with two perfect sets of genitals, worried authorities. Eighteenth-century medical manuals emphasized the legal regulations that applied to hermaphrodites, including laws of marriage, which derived from Jewish Talmudic law and ancient Latin canon and civil law.[29] For example, James Parsons, despite arguing in his 1741 English treatise *A Mechanical Enquiry into the Nature of Hermaphrodites* that human hermaphrodites did not exist, listed each possible legal question, from whether a hermaphrodite should be given a male or female name at birth to whether or not a hermaphrodite should be allowed to marry or divorce. Parsons' answers to the questions required that hermaphrodites or their parents make a permanent choice of sex. Unlike later medical practitioners, Parsons was willing to entrust this vital decision to the individual most concerned. He stated that, "predominancy of sex . . . ought to be regarded; but if the Sexes seem equal, the Choice is left to the Hermaphrodite."[30] Parsons would not, however, have approved of Hall's movements back and forth across the gender divide, for he emphasized choosing one sex.

Although Parsons detailed the legalities relevant to persons with ambiguous genitals and advised such individuals and their parents on the correct course of conduct, he nonetheless denied the existence of human hermaphrodites. He defined a hermaphrodite as "an Animal, in which the two Sexes, Male and Female, ought to appear to be each distinct and perfect, as well with regard to the Structure proper to either, as to the Power of exercising the necessary Offices and Functions of those Parts."[31] Lower forms of animal life, including earthworms, snails, and some reptiles, may display perfect hermaphroditism—entire male and female sexual organs, each with normally functioning sexual and reproductive capability—but not humans. Parsons was right. Individual humans, unlike hermaphroditic earthworms, are not able to reproduce as either sex. Eliminating hermaphroditism as a human phenomenon, however, validated medical and laypeople in their insistence on the rigor of two discrete, mutually exclusive sexual categories, which did not easily encompass all bodies.

Apparently, Hall's body could not be easily classified, and Hall's life reflected that ambiguity. Hall told his history to the court. In England "she" had been baptized Thomasine and until the age of twelve lived with her parents near Newcastle upon Tyne. She spent the next ten years at her aunt's house in London. After her brother became a soldier, Hall dared to cut her hair, wear men's clothes, and join the army. "He" served an unspecified time in the military and then resumed life as Thomasine. According to the court deposition, "Hee changed himselfe into woemans apparel and made bone lace and did other worke with his needle." Not content to remain a woman, Hall decided to adopt a new persona and emigrated to Virginia as a male indentured servant. Once again Hall donned masculine garb. In Virginia, despite his status as a bound laborer, s/he exercised a predilection for crossing back and forth between genders.[32]

When asked "wether hee were man or woeman," Hall answered, "both man and woeman." Hall's own description of his/her genitals suggests that Hall was a hermaphrodite. S/he explained that s/he had features of both sexes and added that s/he "had not the use of the mans parte," though s/he also said there "was a peece of fleshe growing at the . . . belly as bigg as the top of his little finger [an] inch longe." Those who viewed his/her body were uncertain as to which sex Hall belonged, for when a group of female examiners saw this piece of flesh and asked if "that were all hee had," s/he answered, "I have a peece of an hole."[33]

Had the court been able to decide which of Hall's sexual characteristics were predominant, it might have required him/her to assume and maintain this preferred sex. Such a solution would have been consistent with scripture-based laws as interpreted by Talmudic commentaries and consonant with early modern European customs. Instead, the court acknowledged Hall's own self-description as a person embodying both sexes. It decreed that henceforth s/he be required to wear a paradoxical costume consisting of "mans apparel, only his head to be attired in a Coyfe and Crosscloth with an Apron before him."[34] The court did not wish to endorse and promote uncertainty; it chose the sanction, I believe, to preclude future acts of deception, to mark the offender, to warn others against similar abomination, and to reduce the possibility of Hall's sexual coupling. The court's ruling made it impossible for Hall to seduce the unwary of either sex, should s/he attempt to do so, and then to have coitus with the "wrong" sex. This was not a tolerant and understanding ruling permitting Hall to switch between male and female roles as circumstances allowed and opportunities afforded. It prevented any sexual autonomy and ability to blend in with the populace. Hall would have to live the rest of his/her days as a public freak and laughingstock, an ambiguously gendered being, at once male and female.

There was no category of intersex into which the dual-sexed Hall could be fit; there were men and there were women. Hall therefore embodied an impermissible category of gender. Hall might have favored a laissez-faire approach to sexual expression, but the authorities insisted on precise rules of gender display that would reflect and announce his/her equivocal condition. The court's judgment, mandating the simultaneous performance of both genders, rose from the impossibility of clear classification. Ironically, its solution confounded social conventions; individuals did not normally go about in both male and female attire. Though the court might have been less concerned with punishing Hall than with protecting unwary townspeople from sexual congress with a person of the wrong sex, its decision was devastating to Hall's dignity. By this humiliating sentence, Hall was marked as a creature of indeterminate sex, a ludicrously dressed object of disgust, amusement, or pity.[35] Hall could no longer switch between living as a man and living as a woman, nor could Hall live solely as either a man or a woman—only as a public spectacle of no specific gender. Unfortunately for historians, Hall drops from public records after the court's decision. We can only hope that s/he worked off the indenture, changed name and location, eschewed the farcical costume, and resumed life as whichever gender(s) s/he preferred.

Hall would not be the last person in early America to move back and forth between genders. In the colonial period conventional masculinity and femininity were rigidly defined yet nonetheless transgressed, and it may be that the indistinct nature of these people's genitals prompted their shifting between the genders. Although sources are limited, some newspaper evidence depicts what the historian Alfred F. Young has aptly termed "a hidden world of plebeian deception and disguise."[36] The *Pennsylvania Gazette* in 1764 published a story of a woman, Deborah Lewis, who had "constantly appeared in the female Dress" and was always assumed to be a woman. She "suddenly threw off that Garb, and assumed the Habit of a Man." As if to certify that she was truly a man and that hers was not merely a case of cross-dressing, the paper reported that she was "on the Point of Marrying a Widow Woman."[37] In 1770, the *Pennsylvania Gazette* printed another article, presumably about the same Deborah Lewis, suggesting that the Lewis story had become something of an urban legend.[38] The second piece provided details of Deborah's infancy, when supposedly as a baby she bore "a similarity to both Sexes." At her birth there was apparently discussion as to what apparel the baby should wear. It was decided that she be dressed as a female, and she was baptized as such. She "passed for a Woman" for twenty-three years. As an adult, Deborah Lewis lived with a woman who became pregnant and declared Deborah the father. The paper reported that they got married and that Deborah added the man's name Francis, calling himself Deborah Francis Lewis. According to an entry in a book of genealogies, an obituary from 1823 recorded the death of one Francis Lewis, who for thirty-two years "dressed as a woman and was supposed to be such. Afterward he assumed male apparel, married and raised a family."[39]

Was Deborah Lewis a woman who lived as a man, a man who lived as a woman, or an intersexed person whose ambiguous genitals allowed him/her to do what seemed appropriate and natural at different points in life? Without more substantive sources, it is impossible to determine. What will become clear below is that concerns over gender-crossing often involved anxiety about same-sex sexuality. The medical and legal conversation about hermaphrodites, in particular, was often conflated with discussion of same-sex sexuality, especially among women, for hermaphrodites were often thought to be women with long clitorises, capable of and interested in sexual penetration.

Sexual Ambiguity and Same-Sex Intimacy

In 1696, Massachusetts adopted a law against cross-dressing, perhaps to thwart same-sex intimacy or perhaps, troubled by gender masquerading, colonial lawmakers believed that cross-dressing, like homosexuality, belonged in the category of serious offenses. The Hebrew Bible states that a woman or man who wears the clothing of the opposite sex is an abomination to the Lord (Deut 22:5), and it also deplores mingling of any sort; one cannot wear linen mixed with wool; one cannot yoke an ox and a donkey together or sow a field with two kinds of seed (Deut 22:9–11). The Middlesex County Court, in 1692, seemed to be of similar persuasion about the dangerously disordered character of mingling. In charging a woman named Mary Henly with wearing men's clothes, the court contended that those were offenses "seeming to confound the course of nature."[40] In Haverhill, Massachusetts, in 1652, Joseph Davis was convicted of "putting on woemen's apparell and goeinge about from house to house in the nighte."[41] Twenty-five years later, in 1677, Dorothy Hoyt, a woman of Hampton, New Hampshire, was convicted of "putting on man's apparel."[42] Even outside Puritan New England, colonists lived in a world dominated by Christian belief. Women and men had their respective places in the divine scheme, and crossing from one category to the other, to perform what the historian Susan Juster has called "social hermaphroditism," violated providential order.[43]

Unlike laws against homosexuality, which typically punished men but not women (the Bible does not interdict sexual intercourse between women), laws against cross-dressing punished women and men equally.[44] None of these statutes mentioned sexuality directly, making it difficult for modern readers to judge if the threat to the patriarchal order was deemed social or sexual. In other words, did authorities arrest and convict people for cross-dressing because these offenders publicly violated conventional gender roles and biblical law, or was there something left unsaid in this prosecution, namely the fear of same-sex intimacy that might follow from the cross-dressers' deception and seduction of unaware partners? Though no court case I have found specifically linked cross-dressing to homosexuality, it might have been less fraught for judges to punish cross-dressing than to inquire too intimately into sexual matters that were considered heinously unnatural.

In the case of Thomas/Thomasine Hall, Hall's physical sex needed to be established so that his/her sexual behavior could be understood and, if necessary, punished. When asked why "he" dressed as a woman, Hall had responded rather obliquely, "I goe in woemans apparel to get a bitt for my Catt." The historian Mary Beth Norton has suggested that Hall's response may have been a reference to prostitution, an echo of the French phrase, *pour avoir une bite pour mon chat* (to get a penis for my cunt). Hall, an indentured servant, might have dressed as a woman for financial reasons.[45] As a female prostitute, perhaps Hall could find male sexual partners and supplement her meager resources.

But further investigation into Hall's transgression of "wearing women's clothes" revealed more serious issues. The ambiguously gendered servant was rumored to have had sexual relations with a woman, "greate Besse." If Hall was *really* a man, his crime would have been fornication, a common offense in seventeenth-century Virginia. If Hall was "really" a woman, however, the sexual relationship with great Besse might have been considered an unnatural act, or it might have been dismissed as of no consequence, for typically only same-sex liaisons between men were condemned. Norton has argued that the court needed a clear determination of Hall's anomalous sex to determine the nature and severity of the crime: Was Hall a man or a woman? Was the "piece of flesh" that Hall mentioned an enlarged clitoris by which Hall could penetrate other women? Hall's master and other onlookers were interested in Hall's case. These parties were quoted by the court as pursuing the matter so that "hee might be punished for his abuse."[46] The bystanders seem to have been more exercised by their image of Hall's erotic adaptability than by the possibility of ordinary fornication, and their concern highlights the early modern anxiety about same-sex liaisons made possible by the potential fluidity of gender. In the end, the court's punishment of dual embodiment both effectively protected uninformed townspeople from sexual congress with someone of the "wrong sex," and prevented Hall from engaging in same-sex intercourse.

Not only lawmakers and judges but also physicians suspected that hermaphrodites might be tempted to same-sex intimacy. Medical discussions centered on the clitoris, which played an important role in the early modern understanding of hermaphrodites. All the examples Parsons provided in his 1741 treatise on hermaphrodites were, he argued, either truly women with enlarged clitorises or (less frequently) men with small penises, which were hidden in bodily folds and were often accompanied by undescended testicles. Not content with examples from his own observation, his book discusses each case of human "double Nature" that he had encountered in medical literature from the early Greeks onward, proving the descriptions therein mistaken.[47] Parsons joined a long tradition of doctors who examined and discussed hermaphrodites; since the fourteenth century medical men had been interested in the subject and had proffered theories about such occurrences. Some insisted that hermaphrodites were possible. Others, like Parsons, believed that Hermaphroditus, the figure from Greek mythology whose male body was merged by the gods with the female body of the nymph Salmacis, had no counterparts in the real world and it was folly to imagine the existence of such purely mythical beings.[48]

The classification of normal humans as mythical hermaphrodites, according to Parsons, was due to ignorance of human anatomy, particularly of female anatomy. The clitoris, he said, was so little known, so unrecognized as a female organ, it was no wonder that "at the first sight of a large Clitoris, divers odd Conjectures should arise."[49] But medical authors were not completely ignorant of the clitoris. Nicholas Culpeper, herbalist, astrologer, and English translator and coauthor of the 1655 manual, *The Compleat Practice of Physick,* had compared the clitoris to the penis: "It suffers erection and falling as that doth; this is that which causeth Lust in women, and gives delight in Copulation, for without this a woman neither desires Copulation, or hath pleasure in it, or conceives by it."[50] Henry Bracken, author of a 1737 British midwifery manual, also wrote of the clitoral role in sexual pleasure: "The Clitoris, or Penis, of the Woman is erected, which, by its Fullness of Nerves, and exquisite Sense, affords unspeakable Delight."[51]

Most early midwifery manuals found in America, though, offered scant mention of the clitoris and even less of its function. In Dr. Alexander Hamilton's 1790 textbook, *Outlines of the Theory and Practice of Midwifery,* for example, the organ was mentioned on only two of the book's

307 pages. Similarly, William Smellie's 1786 text, *An Abridgement of the Practice of Midwifery*, noted the clitoris only twice, in a list of female anatomical parts. William Cheselden briefly alluded to it in two pages of his 350-page book on human anatomy and barely hints at its significance, explaining in one sentence that it "is a small spongy body, bearing some analogy to the penis in men, but has no urethra."[52] In an 1802 midwifery manual, Thomas Denman wrote, "The clitoris is little concerned in the practice of midwifery, on account of its size and situation."[53]

Though the organ itself was usually ignored or deemed insignificant, clitorises of pronounced size provoked comment and concern. According to Parsons, oversized clitorises were particularly common among African women and could lead to "two Evils: the hindering [of] the Coitus, and Womens abuse of them with each other."[54] Sixteenth-century French medical writers had anticipated Parsons in his latter concern. They suggested that women with large clitorises could give sexual pleasure to other women. One such doctor described the clitoris in 1597 as "that part with which imprudent and lustful women, aroused by a more than brutal passion, abuse one another with vigorous rubbings, when they are called confricatrices."[55]

Even Jane Sharp, who did not hesitate to describe the clitoris's function and form in great detail, linked enlargement of the clitoris both to hermaphrodites and to women's having sex with other women. "Some think," she said, "that hermaphrodites are only women that have their Clitoris greater, and hanging out more than others have, and so shew like a Mans Yard."[56] Implying that an enlarged clitoris could be used for penetration in same-sex relations, she continued, "Commonly it is but a small sprout, lying close hid under the Wings, and not easily felt, yet sometimes it grows so long that it hands forth at the slit like a Yard, and will swell and stand stiff if it be provoked, and some lewd women have endeavoured to use it as men do theirs."[57]

Occasionally, writers combined medical discourse on hermaphrodites with quasi-pornographic tales of women with large clitorises (or other penis substitutes) having sex with each other.[58] *A Treatise on Hermaphrodites*, published in England in 1718, sandwiched salacious stories of women's using their large clitorises for penetrative sex with female partners into an ostensibly scientific account of hermaphrodites. The author, Giles Jacob, introduced the book by explaining the five types of hermaphrodites, a classification he derived from Nicholas Venette's *Conjugal Love; or, The Pleasures of the Marital Bed Considered in Several Lectures on Human Generation.*[59] The first two types look like men, Jacob explained, though their genitalia include "a pretty deep slit between the Seat and the Cod." Both categories are capable of generation. The third type, by contrast, has "no visible privy Parts of Man, only a slit." But these hermaphrodites become men during puberty "through the coming forth of the privy parts . . . in an Instant, and are as valiant in the Adventures of Love as other Males." Jacob cautioned that because these "men" could look very much like women at first, young gentlemen should not be too hasty in their marriages; one could never know whether "in a vigorous Consummation with a very youthful Partner, the imaginary Female should at once appear an Hermaphrodite."[60] The fifth kind "have neither the Use of the one or the other Sex, and have their privy Parts confus'd." These hermaphrodites have the "temper" of both men and women, and their whole constitution is so "inter-mix'd" that it is impossible to say which sex predominates.

Jacob devoted most of his discussion to the type he found most intriguing, the fourth category: "Women who have the Clitoris bigger and longer than others." Jacob knew the function of the clitoris under usual circumstances. He compared it to the penis and agreed with his contemporaries that "without this Part, the fair Sex would neither desire the Embraces of the Males, nor have any Pleasure in them, or Conceive by them." But, just as other European authors had warned, a large clitoris could be problematic, at the very least interfering with heterosexual copulation. "Sometimes the Clitoris will grow out of the body two or three inches" because of the "over much Heat of the Privities," he explained, and will prevent satisfactory intercourse. Though the female's own pleasure may be enhanced by clitoral enlargement, its increased size would prevent a man from "knowing his Wife."[61]

Here the stories of hermaphrodites and lesbians merged.[62] Unable to copulate effectively with men, "robust and lustful" women, "well furnish'd in these Parts," might turn to the "unnatural Pleasures" of sex with women. The middle section of Jacob's book turned to the story of two "masculine-females" of the noble class, Marguereta and Barbarissa, from Italy and France, respectively. Jacob described these two women as "very near equal to the largest siz'd Male" in their faces, shoulders, hands, and feet. Only their hips and breasts were small. A servant spying on their "amorous adventure" observed Marguereta naked and saw "something hang down from her body of a reddish colour, and which was very unusual." Later in the narrative, the focus shifts to the partner, Barbarissa, who was having trouble with the "erection of her female Member," but ultimately succeeded in penetration. Having put such emphasis on their large clitorises, Jacob conceded that both were "suspected to be Hermaphrodites."[63]

As the debate over hermaphrodites' existence continued, the clitoris remained a significant marker. Medical writers commented on its dimensions, increasingly linking its magnitude to homosexual activity between women. In 1807, for example, Dr. William Handy of New York believed that he had seen an actual hermaphrodite, "a person participating of the parts of both sexes." Cognizant of the common understanding of hermaphrodites, he had always assumed that "an animal, uniting the sexes distinctly, had no existence in nature." And so when presented with the "opportunity of visiting and examining so rare a phenomenon" in Lisbon, he eagerly accepted. Like other medical observers, Handy had supposed that the term "hermaphrodite" was reserved for those women "in whom the clitoris was found to be of an uncommon size."[64] Handy believed that the individual he examined, though equipped with a penis, was basically a woman (she had breasts, menstruated and had been pregnant twice) whose sexual desire was toward men. He assured his readers that, despite the penis, the woman did not choose to have intercourse with other women.[65] Later medical writers, analyzing Handy's account, disagreed with him and believed that the supposed penis was "of a cliteroid nature."[66] Either way, Handy pushed the association between large clitorises and sex between women further, suggesting that the clitoris sometimes grew to unnatural proportions *because of* "the morbid effect of frequent lascivious unnatural excitement, as we learn to have occurred in the case of two Nuns at Rome."[67] Not only did Handy believe that most hermaphrodites were really women with large clitorises, but apparently the clitoris in some women grew unusually large as a result of sex with other women. Hermaphrodites, large clitorises, and sex between women were bound together in Handy's account.

According to the historian Sander L. Gilman, Europeans believed that sexual irregularities, especially large clitorises, were particularly common among women in Africa, India, and the Caribbean.[68] Indeed, Ambrose Paré had written about the clitoral excision performed by female diviners of Fez, in North Africa, and subsequent European and American authors continued to project genital and sexual anomalies onto the bodies of women of other races and from other continents.[69] In her section on enlarged clitorises, Jane Sharp pointed out that "In the Indies, and Egypt they are frequent"; and in contrast, she claimed that she had never heard of the problem occurring in England. If there were any Englishwomen afflicted with what Sharp called a "counterfeit Yard," she was sure they would "do what they can for shame to keep it close."[70] Similarly, James Parsons traced his study of hermaphrodites to the exoticized body of an Angolan woman. He wrote that upon the arrival in England of this unnamed African woman, considered by many to be a man, he decided to undertake his project to prove that she was really a woman with a large clitoris and that hermaphrodites did not exist in humans.[71] Ultimately, Parsons supported clitoral excision in such cases. In Asiatic and African countries, he said, "the Women have them most commonly very long," and knowing the trouble they can bring, the people "wisely cut or burn them off while Girls are young."[72]

Clitoral excision would come to be a common, though contested, cure for all sorts of female conditions, including incessant masturbation, nymphomania, syphilis, and hysteria.[73] Genital surgery for "hermaphroditism," including, but not limited to clitoral removal would also become medically accepted.[74] Perhaps because seventeenth- and eighteenth-century medical men believed that hermaphrodites were simply women with distended clitorises, it stood to reason that their solution for this incipiently dangerous "abnormality" would be straightforward removal of the organ.

James Parsons was unusually forward thinking in allowing patients to choose their own sex (and, though he did not say it, their own sexual partners). When later doctors made the choice, they prioritized heterosexuality, to some patients' lifelong grief. If an indeterminately sexed patient, even one who seemed predominantly male, expressed sexual interest in men, for example, doctors advocated surgical intervention to make that person's genitals appear female.

In early America, those with ambiguous genitalia escaped debilitating surgery, yet even before hermaphroditism became a condition triggering medical intervention, early medical and legal authorities had opinions on how people with atypical genitalia should be regarded. When sexual performance, lawsuits, illness, or chance brought their condition to the court's attention, sexual lives were no less scrutinized and publicized than was physical conformation. Sex lives were regulated to the best of the court's ability and, often relegated, with optimum early American propriety, to heterosexual marriage.

Courts were certainly concerned lest the physically dubious enjoy sex with both men and women. They were also suspicious that sexual duality could lead to sexual duplicity—an innocent individual might be seduced into sex with the wrong partner. And they were no less anxious lest the ambiguously sexed, eschewing deception, copulated frankly and openly with their own sex.[75] Though the one true sex of a person such as Thomas/Thomasine Hall was, in early American eyes, known only to God, the legal sentence that required Hall to live as neither man nor woman but as a public burlesque of both was an effort to stifle any sexual expression. For even if the court had imposed a sexual identity on him/her, or if Hall had been willing to choose one sex in perpetuity, what if the court or the defendant chose wrongly? What if the court, in puzzlement, fostered a same-sex alliance, or if Hall, in perverted desire, were able to choose a same-sex mate? Intersex bodies were a source of anxiety about same-sex sexuality, for if God created no true hermaphrodites, then a person with indeterminate genitalia had a definite, though sometimes indiscernible, sex. As time passed, medical (and therefore legal) authorities became more and more certain of their ability to distinguish a person's actual sex or surgically to impose a sexual conformation that suited their prejudices against same-sex unions. To uncover the hidden history of intersex is to expose both early American and later attitudes toward sexual normality and difference. In studying early social response to the challenge of ambiguous genitalia and comparing it to contemporary perspectives, we broaden our understanding of the shifting tensions over gender difference and same-sex sexuality.

From Monsters to Deceivers in the Early Nineteenth Century

Facts reveal'd by Goddard's knife,
Sheds light upon the M.D. strife;
For centuries contended.
That Nature steady in her plan,
Confus'd not sexual forms in man,
Her systems pure intended.

But Carey's life, outre and strange!
Illustrates nature's freaks in change;
Virility affected,———
Devoid of ducts, of glands, and muscle,
Physiologists stare! Their wits bepuzzle,
At wond'rous Facts detected!

IN EACH ERA, doctors interpreted hermaphrodites in a larger cultural context of ideas about women, about men, and about what was normal. In the colonial period, unusual anatomies were seen through the lens of monstrosity. Creatures with very atypical bodies were not human at all, some commentators seemed to suggest, and such horrific births signaled divine punishments of parents' depravity, Satan's intrusion, or unruly maternal imagination. By the late eighteenth and early nineteenth centuries, observers were no longer sure that the anomalies were caused by God's judgment or prenatal influence—so how were they to be regarded? Two related cultural preoccupations colored the anxiety about hermaphrodites in the period of the new republic: worries about racial instability and concerns about deception and fraud. The turn of the nineteenth century marked a transition, as the hermaphrodite shifted from monster to person. Sadly, the assignment of personhood was not positive or even neutral; the hermaphrodite morphed from a merely physically monstrous creature to a repulsive and duplicitous one. Even infant patients, in their innocence, bore the disturbing potential of future duplicity.

In the mid-nineteenth century, as doctors became more professionalized, so too did medical assessments of hermaphroditic conditions. Throughout the century and into the twentieth, doctors continued the debate over whether or not true hermaphrodites existed. Their answers to that question dictated how they viewed and interacted with their patients. Overall, whether doctors asserted or denied hermaphrodites' reality, they tried to determine each patient's true, singular sex with certainty, even though the bodies they saw manifested ambiguity. "Undecided" was the one medical conclusion physicians refused to reach. As we will see in the next chapter, when surgery and hormones became available, the drive for certainty, or at least creation of apparent certainty in the patient's external conformation, became more feasible.

Monsters Abate

In the eighteenth and nineteenth centuries, the most benign term used to designate ambiguously sexed individuals was "hermaphrodite." Other, more subjective, labels were frequently applied as well: "hybrid," "impostor," "unfortunate monstrosity." Doctors used similar derogatory descriptors for genital incongruities. In an 1842 article on malformations of the male sexual organs, for example, one doctor referred to "these mortifying and disgusting imperfections."[1]

The published story of James Carey, who lived in Philadelphia in the 1830s—which still employed such terms as "hermaphroditic monster"— illustrated the emergence of a new sympathy for those born with unusual genitalia. We learn about Carey through the autopsy report written by the celebrated artist James Akin. According to Akin, Carey lived his entire life as a man, but the account describes a creature more beastly than manly. Akin went to great pains to prove the veracity of his narrative, appending signatures by attending doctors confirming the shape of Carey's genitals as well as his general demeanor. Akin recounted Carey's sad and pitiful life for readers. A self-imposed recluse in particular dread of exposure, Carey vigorously guarded his privacy and shunned any interaction with people beyond that necessary to make a living. Akin wrote:

> Conscious that busy intermeddlers might surprise him sleeping, and when in a state of nudity, discover his strange malconformation, he continually girded his pantaloons securely about his loins, and when thus shielded, he would confidently retire to rest, conceiving the vestment a fortress of impregnable strength to protect him against all infractions during repose, . . . or changing his apparel, he would resort to every preventative for guarding against sudden obtrusion, determined to punish with promptitude infringements of the curious, who should violate his sanctuary.[2]

Akin portrayed Carey as a stooped hunchback, "exhibiting features of a grotesque melancholy aspect." His eyes were heavy and dull, his nose flat, "not unlike the lesser Ouran Outang or Pongo Ape of [the eighteenth-century French naturalist the comte de] Buffon." He emitted "preternatural discharges" from his nose, "yielding a horribly foetid stench" that could cause severe nausea or vomiting among onlookers. Provoked by the "very strange and unnatural conformation of his genital organs" and the nasal hemorrhage to which he believed they were connected, Carey took to "venting most vile and impious imprecations, full of profanation and blasphemy, offensively disgusting to a chaste ear, and outraging all decency."[3]

Akin's description, supported by documentation from those who knew Carey and from doctors who witnessed the postmortem, exposed a most distressing and pathetic character. Akin depicted a social recluse. Carey remained chaste throughout his lifetime; indeed, he showed an "incorrigible aversion" to women's "soft blandishments" and never committed, in Akin's words, "debasing earthly drudgery in commerce with the sex."[4] Akin's tone suggested that this was for the best, given Carey's bodily conformation. While nineteenth-century doctors argued whether there were "true" hermaphrodites (Akin claimed that his report would "elucidate if not confirm the hypothesis, that hermaphroditic characters" exist), there lingered in their gradual medicalizing of this condition an older, harsher interpretation of hideous deformity. As the description of James Carey

This lithograph drawing of James Carey as a stagecoach driver emphasizes his stooped, almost monstrous, form. The artist James Akin detailed Carey's "coarse, uncouth, and abrupt manners," as well as his "continued irascibility of temper, often breaking out in peevish paroxysms of passion." Image from James Akin, *Facts Connected with the Life of James Carey, Whose Eccentrick Habits Caused a Post Mortem Examination* (Philadelphia, 1839). Courtesy American Antiquarian Society, Worcester, Massachusetts.

suggests, many doctors and other commentators never entirely abandoned this motif and slipped deprecating words of repugnance into medical descriptions of their patients.[5]

Akin's report nonetheless combines monstrosity with a touch of humanity. Although Carey would sometimes "storm in phrenzied gusts of furi-

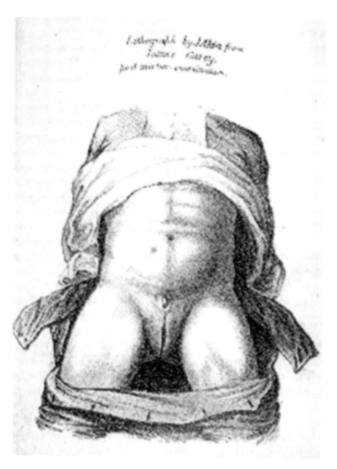

James Akin sketched this image at James Carey's postmortem examination in Philadelphia on June 5, 1838. Akin attested that attending physicians approved his anatomical representations. Image from James Akin, *Facts Connected with the Life of James Carey, Whose Eccentrick Habits Caused a Post Mortem Examination* (Philadelphia, 1839). Courtesy American Antiquarian Society, Worcester, Massachusetts.

ous and ungovernable rage," frightening those who crossed his path, he was able to contain himself at work, Akin wrote, and he displayed honesty, loyalty, and punctuality in his duties. Whether he was working at a foundry or in a factory or driving a stagecoach, at work he exhibited a "firm and manly deportment."[6]

But if Akin recognized that Carey was not all monster, neither was he all man. His voice "partook of a faint boyish squeak." And with his long, lean, and "smooth polished limbs . . . encased in a skin of delicate whiteness," Carey had an "effeminate appearance." Henry G. King, pastor of a Methodist church, attested in an addendum to Akin's account that Carey's skin was "smooth and sallow" and that his chin sported no beard. "In a word," King concluded, "he neither resembled the masculine nor feminine gender."[7]

His body may have been ambiguously gendered, but for some observers, like this minister, Carey's soul humanized him. King, along with employers and landlords, tried to save Carey, especially when he became very ill, exhorting him to prepare to meet God before his death. King's first visit yielded no signs of repentance; at the second visit he saw "some degree of warmth."[8] We will never know if Carey experienced conversion, since he died before the minister could see him again. That Akin chose to include the minister's account of his personal interaction with Carey along with the autopsy report is telling. Carey was no longer just a monstrosity or a dissected body, but a human being with a soul that could achieve forgiveness.[9]

Carey spent his entire life protecting his body's secret, but when he died he could no longer guard his privacy. The coroner "marveled at the novel and singular spectacle" and told others that "all was not right." These words led some to suspect that Carey's death itself was suspicious, and so four days after the burial, the coroner ordered an inquest. Carey's body was exhumed and became available for medical scrutiny, dissection, a published autopsy report, a poem, and subsequent historical analysis.[10] Akin's story of Carey's life bridged the gap between an older conception of hermaphrodites as monsters and a newer emphasis on personhood that combined a description of anatomical difference with moral evaluations of the person's life.

Hermaphroditism Transmutes

By the antebellum period, a new scientific, medical discourse dominated by professional physicians emerged. As scholars of medical history have pointed out, physicians then established themselves as a professional body. American doctors began attending medical schools in greater numbers, and they shared their cases with colleagues in newly instituted medical publications.[11] Doctors wrote about a variety of cases involving unusual genital presentations, including elongated penises, urethral strictures, vaginal fistulas, and hypertrophied clitorises. Even accounts of such cases that did not specifically focus on hermaphroditism offered doctors the opportunity to compare one patient's presentation with another, and their reports often referred to and judged hermaphroditic conditions. In the Surgeon-General's Catalogue Index of published medical articles from the seventeenth to the twentieth centuries (series 1–5), there are over a thousand citations for "hermaphroditism" from the nineteenth century, several hundred of which refer to cases in the United States. Another 2,700 cases are listed under "genito-urinary," and many of them overlap with conditions described elsewhere as hermaphroditic.[12] Many foreign cases were reprinted or referenced in American journals as well, and their inclusion indicates a sharing of knowledge among nineteenth-century practitioners, though such sharing often took the form of challenges and disputes.

The nineteenth-century medical journal articles, not unlike today's medical essays, can be interpreted as conversations among doctors, as they read and rebutted each other's work. Often physicians referenced previously published cases to buttress their own opinions. The cases concerning hermaphroditism were no exception. Doctors cited sixteenth-century precedents, for example, in order to argue to their colleagues that hermaphrodites did not exist, despite the cases of ambiguous sex that they continued to see. The patients they evaluated, then, were "really" men or women, and doctors published their unusual cases to prove their points and validate their medical authority.

Nineteenth-century doctors, had they been given the opportunity, would never have forced the seventeenth-century Thomas/Thomasine Hall to live as simultaneously male and female. They were far more intent on deciding the patient's sex definitively and in making sure that the person lived by their professional assessment. In their zeal to achieve sexual certainty, doctors did not hesitate to judge their patients' gender performances as mistaken, if not deliberately fraudulent.

The Possibility of Fraud

The possibility of swindle and deceit looms large in nineteenth-century discussions of hermaphroditism. Certainly crossing the gender divide, as Hall had done, was risky. Rumors might spread, and the suspicion that one was not who one claimed to be, or who one was assumed to be, could be hazardous. Scholars have asserted that the 1830s and 1840s saw a particular anxiety about middle-class respectability and the dangers of deception.[13] At a time of increasing geographic mobility and urbanization and new impersonal commercial networks, there were new opportunities to remake one's self and perhaps to deceive others. The "self-made man" might be a confidence man or—shockingly—even a woman. In this mutable world, changing one's gender was the ultimate dishonesty.

In 1840, for example, the *Boston Medical and Surgical Journal* published an article about a purported hermaphrodite who had lived some years as male and some as female. The author, an anonymous physician, described the subject's ambiguous features: long, black hair arranged in a "feminine mode," a face with "masculine coarseness," but with "a feminine complexion," facial hair like a man, but earrings like a woman. It was said that s/he engaged in intercourse as a person of either sex. Despite the ambiguity of indicators, and the subject's presentation as female, the doctor expressed no doubt about his subject's sex and portrayed "him" as unequivocally disingenuous, as a "case of imposture."[14]

What are we to make of this perplexing person? Was she a woman dressed in men's clothes, as so many supposed and she insisted? Was she truly a man, equipped with "male organs entire," as two doctors observed? Did the "piece of dead flesh" she referred to on her body make her female or male? Why did she variously live life as a woman or a man? Why did the doctor feel entitled to pronounce her male, even as she presented herself as female?[15]

In many ways, this case typified mid-nineteenth-century medical discourse on hermaphrodites; the themes of dishonesty and sexual promiscuity lurk in what are otherwise dispassionate and clinical medical cases. Foremost in the narrative is the report of the subject's shiftiness, as if bodily ambiguity meant that the person's word also lacked clarity and could not be trusted. The doctor detected "imposture" and continually referred to the subject as male. Nonetheless, the doctor noticed that the subject's style of walking was so feminine that "no one could avoid the suspicion that the individual was a woman in male attire." In fact, that suspicion had brought public notice and led to her arrest for being a female disguised in men's clothing.[16] From the jail, she was sent to the almshouse, where the article's author was requested to help the superintendent with the examination. Stories had apparently spread about the prisoner. She was known as a hermaphrodite, and the doctor learned "that rumors circulated of his performing the copulative functions of either sex."[17]

The examining doctor used the masculine pronoun "he" throughout his account, thus making his own evaluation of the case evident, although the prisoner's words, as much as they can be ascertained from the description, suggest that she identified herself as a woman. "He" was bashful, according to the doctor, and did not want to submit to a medical examination because "he" said he "was menstruating." The doctor's inability to accept the patient as female, despite her own wishes, led to incongruities, the very ambiguity (a man menstruating, for example) that the doctor was trying to avoid by categorizing the patient as male. The superintendent tried to persuade the patient through bribery and threats to submit to either the doctor's or the superintendent's examination, but to no avail. Finally, she acceded to an exam by a female inmate.

The woman inspector declared that "the female organs predominated," and so the patient (still called "he" by the doctor) settled into one of the female wards. She maintained that she had been raised female and that she had worked as a kitchen maid and as a female circus performer. While at the almshouse she worked in traditionally female jobs, such as washing clothes. She said that she had no "inclination" toward either sex romantically. Regarding her genitalia, she referred to herself as having "a piece of dead flesh hanging down."[18] She worked for a few weeks at the almshouse but then contracted pneumonia and died.[19]

The doctor, with a colleague, eagerly performed an autopsy, confirming his own suspicions that the deceased was not a woman. He "found male organs entire and well developed, and no semblance whatever of those of the female." The other doctor present at the autopsy recognized the deceased as a person who had once appealed to him for a certificate authenticating his male organs. Apparently, having been repeatedly harassed for being a woman dressed in men's clothes, she sought to prove via a doctor's certificate that she did indeed possess male genitalia. She had told the magistrates before whom she appeared that she dressed as a man in order "to avoid the importunities of the sex." In other words, too often sexually harassed as a woman, s/he dressed as a man.[20] The subject was apparently adept at procuring funds as either sex. While living as a man, s/he successfully convinced "benevolent ladies" that s/he did so to avoid the male advances so commonly made to women. They took pity, offered sympathy, and contributed to his/her financial aid. Were the "benevolent ladies" the victims of a con artist? The doctor's written evaluation of the case encourages this interpretation; readers are left with the impression that his most important diagnostic tool was his persistent desire to uncover deceit.

Given the pervasive suspicion of supposed hermaphrodites found in many medical accounts, it is perhaps unsurprising that the descriptions betray white middle-class anxieties about the boundaries of both class and race, as we shall see.[21] Statements in the medical narratives that subjects were poor or in an almshouse, for example, often suggested that either laziness or dishonesty had landed them there. One patient suffering from a malformation of his urethra was introduced to journal readers as a famous pickpocket originally from Holland, who spent many years of his life in the New York state prison. He was said to have exaggerated his symptoms (pain in the kidneys and scant urinary void) in order to acquire greater medical benefits from his native country.[22]

Similarly, in 1829 a Boston medical journal reported the case of Mary Cannon, a woman admitted to the charity ward of Guy's Hospital in London who had lived many years of her life as female and many as male. She held numerous working-class jobs (toiling in a brickyard and as a milkman, greengrocer, and maid), and the doctor noted that "her habits and manners were rude and bold, sometimes indicating a degree of derangement." "Suspicion hung about her," the doctor noted, particularly concerning her sexual proclivities; indeed, while she was a maid the other female servants never accepted her, and "care was always taken to provide her a separate bed."[23]

If there were no such beings as hermaphrodites, as leading authorities insisted, then how did they classify the patients they encountered? According to John North, author of a two-part article on hermaphroditism and other "monstrosities" published in America in 1840, "All the alleged cases have merely arisen from various malformations, either of

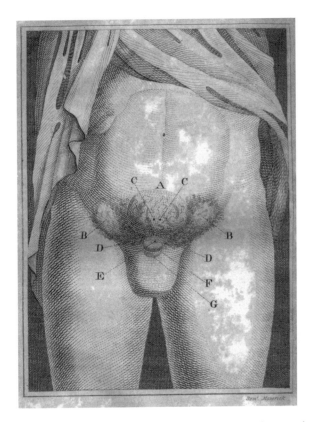

The text accompanying this picture of a New York "pickpocket" in the *New York Medical and Physical Journal* remarked that the subject, a man named Robinson, was thought to have exaggerated his symptoms of kidney pain and scanty urine discharge in order to receive medical benefits from his native Holland. This picture is typical of medical journal illustrations. Image from the *New York Medical and Physical Journal,* July-September 1826. Published with permission of ProQuest Information and Learning Company.

imperfectly formed men or imperfectly formed women. In consequence of malformations of the genital organs, girls have been educated for boys, and vice versa." North observed that the "mistakes" were sometimes discovered quite late in life, pointing out that individuals who have been raised as males "suddenly astonished the world by being delivered of children; and in other cases, malformed men, who have been supposed to be women, have lived for many years as the wives of men, and the fact has not been discovered till death took place."[24]

North's explanation suggested the possibility of deception, whether intentionally perpetrated or not. The following case illustrates the consequences of fraud and the rewards of male privilege. What was seemingly easy to ascertain turned out to be quite debatable. In March 1843, Dr. William James Barry of Hartford, Connecticut, examined a twenty-three-year-old man, Levi Suydam, of nearby Salisbury. Apparently, the Whigs of Salisbury had presented Suydam to the town's board of selectmen to be acknowledged as a freeman and property holder able to vote in the upcoming election. (Connecticut had a property requirement for voting until 1845.) As the election was closely contested, Suydam's petition was challenged. The opposing party did not dispute that Suydam owned sufficient property; it objected on the grounds that "he was more a female than a male, and that, in his physical organization, he partook of both sexes." Women were, of course, unqualified to vote—both legally and (it was believed) biologically.[25]

The close physical scrutiny that Suydam endured recalls the examinations imposed on Thomas/Thomasine Hall in the seventeenth century, except that, consistent with the professionalization of medicine, the experts were now town physicians rather than midwives or community leaders. The first intimate inspection of Suydam revealed that he was a man and thus eligible to vote. The doctor found a penis, an underdeveloped scrotum, and one small testicle. Dr. William Barry pronounced him a "male citizen, and consequently entitled to all the privileges of a freeman." But Dr. Barry's examination was not the last. Later, on election day, a Dr. Ticknor challenged Suydam's admission as he came forward to vote. Dr. Barry intervened, explained his findings, and invited Dr. Ticknor to examine Suydam privately. Dr. Ticknor concurred with Dr. Barry, confirmed that Suydam was a man, and so reported to the selectmen. Identified as a male by the presence of male genitals, however diminished, and authorized by two doctors, Suydam voted, and according to Dr. Barry's account, the Whigs, who had presented Suydam initially, won by only one vote.[26]

A few days after the election, an apparent ruse was revealed. It was discovered that Suydam bled monthly. According to Suydam's sister, a Mrs. Ayers, who regularly washed his clothes, there was no doubt that Suydam menstruated, though not as heavily as most women. Dr. Barry requested a meeting with both Suydam and Dr. Ticknor to reevaluate the situation. Suydam admitted that he bled and again submitted to an inspection. This time, Dr. Barry's examination was far more extensive, assessing both Suydam's physical attributes and his social qualities.

Likely influenced by his knowledge of the bleeding, Dr. Barry now became impressed with Suydam's "feminine" features. Barry drew attention to the fact that Suydam was only five feet two inches tall, had light hair, a fair complexion, and, notably, a beardless chin. He also had narrow shoulders, wide hips, and "well developed mammae, with nipples and areola." Somehow, these signs of womanliness had escaped Dr. Barry's notice at earlier examinations, when he apparently checked only the genitalia. Dr. Barry now deduced from these secondary physical attributes that, despite the presence of the male organs, Suydam exhibited "in short, every way of a feminine figure." Even Suydam's genitalia looked different on second inspection. Dr. Barry inserted a female catheter into the opening through which Suydam urinated. The catheter entered a space (not his penis), which Dr. Barry now understood to be "similar to the vagina, three or four inches in depth." Dr. Barry inquired further and discovered that at Suydam's birth a doctor had surgically made this opening.[27]

Suydam had amorous desires for men and, according to many observers, "an aversion for bodily labor, and an inability to perform the same." Others had noticed his "feminine propensities," which included a "fondness for gay colors, for pieces of calico, comparing and placing them together." By the conclusion of Dr. Barry's account, there seemed to be no reason to doubt that, although Suydam perhaps "partook of both sexes" and although he had voted as a man, he leaned toward womanhood both physically and socially and was more female than male.[28] Suydam's earlier efforts at maleness could be interpreted as duplicitous, leading to the worst kind of voter fraud. Detecting and exposing such deception was exactly what doctors wanted their science to accomplish.

Yet if many wished for an ideal clarity and definitude, the real world of the nineteenth-century United States was muddy and slippery, and Americans exhibited a paradoxical attitude toward confidence men, impostors, shady entrepreneurs, religious frauds, and cross-dressing women. Though such individuals were conventionally deplored, they were also admired as tricksters who got the better of the establishment, and tales of their supposed exploits were eagerly circulated. Among the most popular accounts of women passing as men was an often republished novel, *The Female*

Marine, about a woman pretending to be a man, written by a man pretending to be a woman. In the book, Lucy Brewer goes off to fight in the War of 1812, serving as a sailor aboard the USS *Constitution.* She displays resourcefulness, self-reliance, and mobility—characteristics commonly deemed male but that this female marine appropriates to deal with her extraordinary predicament. After scenes of danger, suspense, and near discovery, Lucy returns to acceptable female dress and sensibility and marries an appropriate suitor, whom she had met during her masquerade. All's well that ends well in *The Female Marine,* as characters revert to their true natures, aligned with prescribed categories of gender and sex. The chaotic world of gender impersonation settles into one of blissful morality, and Lucy accepts the conventions of the cult of true womanhood.[29]

Sex and Race

Embedded in many of the medical accounts were worries going beyond the threat of dishonesty and illicit sexual relations to the far more troubling threat of inexplicable sexual transformation. As if individuals shifting back and forth between the genders at will was not bad enough, doctors reported startling cases of people who suddenly and involuntarily changed their sex. In 1850 a Boston medical journal reported the case of a fourteen-year-old who had been born female and christened Rebecca, but whose body altered into that of a boy. His father successfully petitioned the court to have the child's name changed to William. Similarly, the *Medical Examiner* in 1839 reported the mysterious masculinization of an eighteen-year-old woman. At her birth, there had been some doubt as to the baby's sex but, according to the author, the "gossiping females" present decided that her organs of generation looked more female than male. Although as a child she engaged in "manly sports and the labours of the field," she wore female clothing and lived as Elizabeth. When she turned eighteen, however, her body changed. She was nearly six feet tall and had begun to grow a beard. Old enough to make her own decisions about gender and sexuality, she abandoned her old name and female identity, lived as a man, and married a woman.[30]

The unnerving possibility that individuals could suddenly change sex paralleled the early national preoccupation with race, racial categories, and the possibility of changing racial identity. In the early republic, as

Americans sought to find social and political order in their unsettled national life, the potential transmutation of race raised serious questions. In 1796, Henry Moss gained wide celebrity as an American-born man of African descent who in middle age somehow turned white. Another man, James, from Charles County, Maryland, lived for fifteen years "a black or very dark mulatto colour," and then white spots began to appear. Gradually increasing, the spots grew until he had become entirely white, except for a few lingering dark spots on his cheekbones. His child, the article duly noted, was born with white spots, suggesting that the offspring's transformation might come soon as well.[31]

Discussions of racial mutation were often linked to issues of paternity, as physicians and lawyers questioned whether or not a white woman's dark-skinned child was fathered by her white husband or by an African American. In 1845, the *Western Journal of Medicine and Surgery* reported a case in which a wealthy planter's wife gave birth to a child so dark that "unpleasant suspicions were awakened among her acquaintants." The woman's husband died shortly after the birth, and the woman brought a slander suit against a physician who accused her of giving birth "to a mulatto child." Nine doctors testified at her trial as to whether a white man could have fathered a child with such dark complexion. The description of the boy's bifurcated body sounds remarkably like those hermaphrodite accounts of bodies where the upper half looked female and the lower part looked male, or vice versa. This child's torso showed two different shades of color, "the chest and axilla being nearly white, while the abdomen is dark, the change occurring abruptly and being marked by a well defined line." The end of the boy's penis, experts noted, was "quite blue," while the rest of his genitals were "of the complexion of the general surface." Surprisingly, the child was changing color, just as Henry Moss had, "growing gradually whiter since birth."[32]

At the trial experts proposed several possible reasons for the boy's unusual appearance, including an explanation largely discredited, though nonetheless vigorously defended, by the mid-nineteenth century: his mother's imagination. The court first explained that though the boy's father's family hailed from Germany and was quite fair, the rest of his family had descended from dark-skinned Gypsies. The boy's mother and maternal uncles also had dark complexions. If that were not enough rationalization for his skin tone, testimony included the evocative detail that during pregnancy, the mother had been "repeatedly frightened by reports of negro insurrections." In the end, the defendant won, proving that the woman had "not been above suspicion," and that she was likely guilty of "criminal connection" with her carriage driver.[33]

Larger questions about race, skin color, and heredity arose at the trial: "Is not a mulatto from a white woman darker than one from a black woman?" "Is it not unusual for mulattoes to grow darker instead of white?" "Do you know of any mode of bleaching by which the skin might be rendered white?" "Do you think it possible for color to show itself in the third and fourth generation?"[34] The boundaries between black and white were not nearly as rigid as some might have wished to believe, particularly between generations. This genre of cases raised the alarming possibility that sometimes changes occurred visibly on people's bodies, making them darker or lighter than they had been previously.

The press reported disturbing accounts in which whites were rendered black—as when light skin darkened among white victims of yellow fever, a disease that spread through eastern cities in the 1790s. Or when white U.S. sailors, captured by Barbary pirates, risked reductions to servility and, some feared, actual physical blackening in North African prisons. One case recorded an elderly white woman's skin turning entirely black for over a year, evidently as a result of her grief at the violent death of her daughter and grandchildren. Doctors described "black matter" even in the skin of whites, suggesting that, although "it seldom discoulours our [white] skin," the prospect of such transmutation existed, as the elderly woman's case proved. With Southern manumission, emancipation in the North, and the growth of the free black population, opportunities for racial mixing inspired intense white efforts to stiffen measures that controlled and separated African Americans.[35]

Henry Moss's legendary case remained in circulation among nineteenth-century writers as they disputed whether "negroes" and whites formed two distinct races or variations of one race. In the late 1850s, a vehement debate raged in the pages of the *Medical and Surgical Reporter* on the nature of species division and the possibility of interbreeding between different species. The series began with a provocative article by a physician, W. S. Forwood of Maryland, whose two-part treatise, arguing that blacks were, in fact, distinct from (and inferior to) whites drew much response. Forwood maintained that the different races of humans were species, that among them "the white and black species . . . are undoubtedly the most remote from each other," and that skin color distinguished the species.[36]

In this context Henry Moss's condition was proffered. How did skin arrive at its color? What would cause it to change color? The unspoken subtext troubled the authors and no doubt readers as well: could people change their skin color and hence their race? Moss's example was used to prove that though some parts of the skin could turn from black to white, not all the skin (or hair) would turn. Those parts of the skin exposed to the sun would stay dark much longer, according to one pseudonymous author, Senex. Senex advocated the unity of the human species rather than Forwood's plurality of species and argued, as many others had, that natural causes, specifically an intensely hot climate, not inherent inferiority, were responsible for Africans' color. Attributing diversity within the species to environmental influences, Senex raised Moss's case to prove that, even with the lightening of his skin, Moss did not change color (or his race) completely, because such changes take thousands of years. Senex, writing in 1857, quoted Samuel Stanhope Smith, moral philosopher, seventh president of Princeton University, and author of an influential 1787 essay, *An Essay on the Causes of the Variety of Complexion and Figure in the Human Species:* "When any dark color has been contracted by the human skin, the solar influence alone, and the free contact of the external air, will be sufficient to continue it a long time, even in those climates which are favorable to the fair complexion."[37]

Ironically, though Senex used Moss's incomplete transformation to show that environmental influences were pervasive and durable, he may have unwittingly bolstered his opponent's point of view. Forwood, too, believed that conversion between races was impossible: since different species do not change into one another, blacks could not turn into whites, or vice versa. People such as Henry Moss may have had a medical condition causing skin anomalies, but "permanence of type" within particular species prevented "any essential organic change of conformation."[38] Forwood marshaled evidence to prove that "species do not change."[39]

Entrenched in discussions of biology and race were suspicions that African Americans were disproportionately affected by genital anomalies, especially elongated penises and enlarged clitorises and labia.[40] In a popular marriage guide from 1850, Frederick Hollick wrote about deformities of the penis that might affect marital relations. In some men the interior of the corpus cavernosum (erectile tissue within the penis) became ossified, "so that a distinct bone always existed in the middle of the organ . . . This is often the case in Negroes," Hollick reported, and "in some of the lower animals it is natural."[41] Accounts of more unusual African American genital anomalies circulated and reappeared. Francis Wharton and Moreton Stillé included in their nineteenth-century textbook a case from 1744 about a clitoris "two inches long and as thick as a thumb, in a negress twenty-four years old."[42] J. W. Heustis wrote of a "negro child six years of age with a lusus [abnormality] of the following character."[43] He described the child's anatomy, which looked female (with "pendulous" labia) upon first inspection but featured a "malformed penis springing from the usual place, but suddenly turning backwards on itself at an acute angle." Convinced—that the child was "neither male nor female," Heustis speculated that the patient he described would never be able to copulate.[44] S. B. Harris's description of an eighteen-year-old slave, Ned, linked an earlier notion of monstrosity to racist stereotypes of African American sexuality. Referring to Ned as "a monster of this singular character," Harris described his "well developed protuberan mammae" and his "large, prominent" pubis, though with a small penis. Harris mentioned Ned's sexual desire for women—he had "strong salacious propensities"—but wondered whether "his amorous advances to the dusky maidens around him" could ever have been consummated.[45]

In the nineteenth century, as Northern blacks figuratively turned white through emancipation, women threatened to turn into men—again, figuratively—as they claimed the political rights of citizenship reserved for white males. One response by those most fearful of chaos and most committed to established structures of power was more stringent classification of such categories as race and sex, based on the conviction that such divisions were embodied and essential. African Americans were thus defined as inferior and servile in their very essence, rendered so not by circumstance, but by nature itself, and women were judged unequal to the task of public citizenship—the domain of white males—because of the essentially dependent nature of their sex.[46]

The (Im)possibility of Perfect Hermaphrodites

Significantly, though medical men wanted to ensure the specificity and stability of each person's sex, uncertainties regarding the criteria for femaleness or maleness abounded, just as they did regarding racial criteria. Were there people whose sex could not be firmly established? Did hermaphrodites actually exist? Late eighteenth- and nineteenth-century medical writers echoed James Parsons' doubts about the existence of hermaphrodites among humans. Yet many writers tried to have it both ways. "Perfect" hermaphrodites *could* exist in the human species, some insisted, but they were extremely rare; most so-called hermaphrodites were, as Parsons presumed, women with elongated clitorises erroneously judged to be penises.[47]

Samuel Farr, author of a 1787 medical jurisprudence textbook published in the United States in 1819, defined "perfect" hermaphrodites as those "partaking of the distinguishing marks of both sexes, with a power of enjoyment from each." Whereas Parsons had required a perfect complement of parts able to "exercise[e] the necessary Offices and Functions of those Parts," Farr pushed the definition a crucial step further by adding "a power of enjoyment." Hermaphrodites had to have both sets of organs *and* be able to use both for sexual satisfaction. Could a hermaphrodite derive sexual pleasure as both a female and a male? That seemed impossible to most writers, though they revisited the question repeatedly throughout the eighteenth and nineteenth centuries. More than one nineteenth-century commentator also wondered if hermaphrodites would be able to impregnate themselves. But medical men were sure the answer was no. No such instance had ever been found.[48]

By the mid-nineteenth century, some American doctors had become familiar with the sorting and stabilizing efforts of Isidore Geoffroy Saint-Hilaire, author of *Histoire des anomalies de l'organization,* published in 1832–36 and soon excerpted in English in Theodric Romeyn Beck and John B. Beck's prominent medical jurisprudence textbook of 1838.[49] Geoffroy Saint-Hilaire argued that hermaphroditism, per se, did not exist; as Beck translated the passage, "the external organs (as a penis and clitoris) have never been found perfectly double." Geoffroy Saint-Hilaire believed that anatomists had resolved the debate about hermaphroditism once and for all; he concluded that "it is anatomically and physiologically impossible." Echoing Geoffroy Saint-Hilaire, author John North denied the existence of perfect hermaphrodites. "Although we see many instances of true hermaphroditism in the animal and vegetable kingdoms," he remarked, "no such cases have ever existed in the human subject; no human hermaphrodite, in the proper sense of the term, has ever existed; not a single so-called hermaphrodite in man has even been capable of performing the sexual functions of both sexes."[50]

But despite assertions to the contrary, some physicians were convinced that the people they saw were indeed hermaphrodites, though they sometimes tempered their assertions by labeling a person a "spurious" or "pseudohermaphrodite" if their external genitalia did not align with their internal anatomy. In 1850 Dr. Jonathan Neill, a professor of anatomy at the University of Pennsylvania, presented a subject he believed was "certainly entitled to the term hermaphrodite." The body of the deceased person arrived at the anatomical rooms of the university for an autopsy. Not much was known about the subject, other than that she had "resided among the degraded blacks in the lower portion of the city," according to the coroner's jury, and that she died of "drunkenness and exposure." The jury was able to surmise from her teeth and "general appearance" that she was between twenty-five and thirty years old when she died. Though the subject dressed in women's clothes, Dr. Neill was not entirely persuaded that she was female. She had large breasts and no hair on her face, two markers that typically would indicate femaleness. But other secondary sex characteristics suggested maleness. Neill wrote that if one looked only at the ratio of the broad shoulders to the narrow hips and also at the shape of the limbs, it would "have indicated the male sex."[51]

What were the definitive markers of sex, then? Did large breasts and a smooth face trump narrow hips and broad shoulders? Upon inspection, the genitalia revealed a similar ambiguity. Dr. Neill said that from a "superficial view" of her genitals, "almost any one would have pronounced the subject to have been a hypospadic male." Since this person was dead, Dr. Neill could go beyond the superficial and perform an autopsy to search for other clues. Inside the body, Neill discovered female internal reproductive organs: uterus, fallopian tubes, small ovaries, and a narrow vagina "of the proper length." Now considering the subject more female than male, Neill classified her condition as "spurious hermaphroditism in the female."[52]

Five years later, the doctors Francis Wharton and Moreton Stillé mentioned the case in their textbook, *A Treatise on Medical Jurisprudence*. What Neill thought most would have seen as a hypospadic penis, Wharton and Stillé described as both a long clitoris *and* a penis. In an unusual display of uncertainty, in one sentence the two described the subject as having a clitoris five inches long and in the next as having a penis.[53] Perhaps Wharton and Stillé were puzzled by the drawing of the subject included alongside Neill's article. As Neill had written, the subject's picture revealed an athletic male physique: broad shoulders, narrow hips, and muscular limbs. She had breasts and a smooth face, and the genitals looked more male than female, as Neill originally suggested. A second drawing depicted a closer view of the penis/clitoris with a vaginal opening underneath. Perhaps Dr. Neill was right initially to conclude that the person was "certainly entitled to the term hermaphrodite." The visual portrayal of the subject is as confusing as the genitals; the coroner had said that the subject died of drunkenness and exposure and lived "among the degraded blacks in the lower portion of the city."[54] The figure here looks more like a Greek god/goddess than someone devastated by alcohol and poverty. Furthermore, as the subject was dead when she arrived at the university, the image of her posing elegantly, standing against a table, adds to the general uncertainty of the case.

Wharton and Stillé recognized that the parameters of the definition of hermaphroditism were shifting, though their language betrays a continued commitment to the monster motif. "Hermaphroditism" was no longer used only to describe the "perfect" union of male and female organs in one individual, since that combination, along with self-impregnation, seemed impossible. By the mid-nineteenth century, the term was used more broadly "for all those cases in which doubts exist concerning the real sex, in consequence of some aberration from the normal type of the genital organs." They admitted that it was often difficult to settle on a person's sex, particularly a child's; sometimes it was difficult even after a mature person had died, as in the case Neill had described.[55] "We can only hope to approximate to the truth, by observing whether there is not some regularity in the freaks of nature, and thus discover, if possible, some

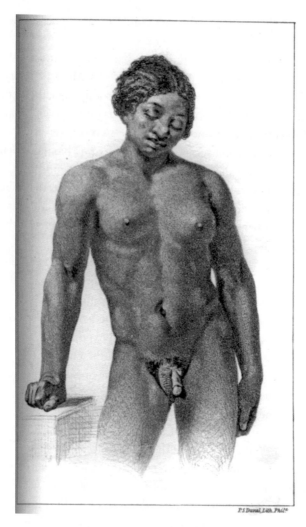

P.S. Duval, Lith. Philª

Physicians were puzzled by this person who had female breasts and a mixture of male and female genitals. She arrived in the examination room dead from alcohol abuse but appears almost regal in this portrayal. Image from Dr. John Neill, "Case of Hermaphroditism," *Summary of the Transactions of the College of Physicians of Philadelphia* 1 (1850). Courtesy Yale Medical School.

uniform correspondence between the visible deviations and those which are hidden from our view." Urging restraint, Wharton and Stillé cautioned against a hasty pronunciation of a person's "true" sex, a prudence that went largely unheeded, as we shall see.[56]

The Significance of Marriage

In deciding the sex of their patients, doctors sought happy endings, hoping to see their patients embrace at least one element of womanhood or manhood: marriage. Physicians first attempted interventionist surgery on genitalia in the hope of making those organs serve the doctors' perception of patients' sexual and marital requirements. One case, in 1833, involved

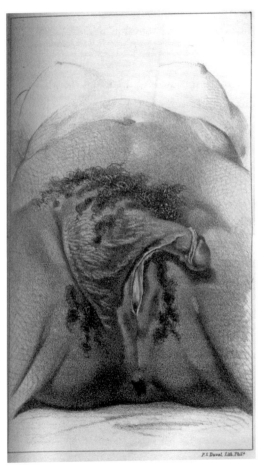

P.S Duval, Lith. Phil.ª

In this drawing the genitals show what physicians thought was either a penis or a clitoris as well as a vaginal opening underneath. Image from Dr. John Neill, "Case of Hermaphroditism," *Summary of the Transactions of the College of Physicians of Philadelphia* 1 (1850). Courtesy Yale Medical School.

a twenty-three-year-old woman who arrived in Dr. John C. Warren's Boston office with a "natural malformation of the generative organs."[57] She had no vagina and requested the creation of an artificial passage. We do not know why she wanted the surgery. Warren's description of her hints at normative heterosexuality. He mentions that she was "well-constituted" with normal breasts and clitoris, but no vagina or uterus could be found. Making an aperture just in front of her rectum, the doctor was able to create a vaginal opening three inches deep and wide enough to admit a finger. After days of profuse bleeding, fever, and pain, followed by dilation of the opening, the wound healed and seemed to remain open. In fact, the doctor reported that "*something like* labia" formed. At her next appointment he noticed "a sanguineous discharge *resembling* the catamenia [menses]," and he thought he could distinguish "*something like* an uterus." The report of the case ends there, with the surgical construction of a (normal) woman who bled, could be penetrated, and, the doctor suggested, could bear children.[58]

The impulse to ensure a patient's future marital prospects is obvious in the following case, though this doctor's interventions were harshly criticized. In 1849, a three-year-old came under the care of Dr. Samuel D. Gross, a major figure in American surgery, author of several influential books on the subject, and a professor at the University of Louisville. For the first two years of her life, the patient had been "regarded as a girl," but at the age of two a strange metamorphosis began. Gradually she started to "evince the tastes, disposition, and feelings of the other sex." In other words, she began to reject her dolls and "became fond of boyish sports." The doctor commented that in every other way this girl was healthy: she had long dark hair, dark eyes, perfectly formed hips and chest, arms and legs, and a lovely face. But upon closer examination, the doctor found what was likely the root of her masculine predilections: ambiguous genitals.[59]

Dr. Gross located neither a penis nor a vagina. Expecting a penis because of the girl's propensity for boyish sports, he discovered instead what he took to be a small clitoris, a "cul-de-sac" instead of a vagina, and, growing inside the labia, what he believed to be one testis on each side. If her testicles had been allowed to mature to puberty, Dr. Gross speculated, they might spur masculine sexual desire, which could lead to a "matrimonial connection." And since the marriage could not be consummated by way of penetration, Dr. Gross thought it best that the testicles be surgically removed. The case, then, turned, not on the little girl's present propensity for "boyish sports," but on her future marital prospects. After the operation, Dr. Gross happily reported, her "disposition and habits" returned to those of a girl, and she took "great delight in sewing and housework" rather than "riding sticks and other boyish exercises." He had seen the girl several times because she lived in his neighborhood and she appeared to be developing normally. In fact, she had an "uncommonly active" mind for a child her age.[60]

For Dr. Gross the single most important justification for the surgery was his young patient's future marital prospects. He could not bear the thought that she would be unmarriageable. The doctor assumed that when the testicles matured, they would arouse a sexual interest, and the girl would seek satisfaction. But with whom? Sex with a man would be unachievable, the doctor implied, because without a proper vagina, penile penetration would not be possible. He mentioned that impregnation was similarly unfeasible. Might Dr. Gross have worried that the testicles the girl possessed could lead her to pursue a female sexual partner? A small clitoris, even if incapable of performing penetration, combined with testicles, might so incline her, Gross hinted, arguing that maturation of her testicles would "ultimately lead to the ruin of her character and peace of mind." A man with no testicles would have no incentive to marry and would be "doom[ed] to everlasting celibacy," according to Gross's reasoning, but a woman with no sexual desire would still be marriageable.[61] Better a woman with no sexual desire than a man unfulfilled, the doctor seemed to suggest.

Dr. Gross believed he made the right decision for this patient, whom he perceived as not "a boy or a girl, but a neuter." He hoped to ensure that she would not be "forever debarred from the joys and pleasures of married life, an outcast from society, hated and despised, and reviled and persecuted by the world." Gross considered his surgery a success, and he published his account in a leading medical journal. He claimed no regrets about performing the surgery and three years later maintained that he had done what was best for the girl and her parents. Dr. Gross recognized the radical nature of his medical intervention, but despite finding no precedent for it, he defended his operation, stating that it was "perfectly just and proper, and vindicated upon every principle of science and humanity."[62]

Not everyone agreed. Whereas other practitioners might have concurred with Gross in considering "dreadful" the "defective organization of the external genitals," the journal's editor found the surgery barbaric. In a postscript to Dr. Gross's account, the editor sarcastically quipped that doctors might just as well administer prussic acid (used as an insecticide and a rat poison) to those afflicted with malignancy, so preposterous, dangerous, and potentially deadly was the intervention. That the editor singled out Gross's operation is telling; most surgery at that time was horrendous, as doctors typically operated without anesthesia, could not control infection, and lost many surgical patients. Wharton and Stillé, in their medical jurisprudence text, were similarly horrified. They held that such an operation "removes merely the *external*." In a response more characteristic of today's intersex activists, they urged readers to avoid surgery, arguing that "it does not necessarily extinguish the sexual instinct, nor deprive the person of 'his only incentive to matrimony,' and finally, in no way relieves him from the odium or aversion with which the malevolent or ignorant may regard him."[63] Yet Gross had his supporters. In an article on morphology of the sexual organs, one doctor suggested castration for "persons with organs so imperfectly formed." For the alleviations of "the sufferings to be apprehended from ungratified sexual desires . . . and for the timely prevention of such sufferings," the author approvingly cited Gross's surgical removal of his patient's testicles.[64]

Perhaps to avoid the surgical and social choices that Gross confronted, medical men recommended raising children with ambiguous genitalia as boys. According to Dr. T. Holmes, "When doubt exists as to the sex of a child, it appears more prudent to bring it up as male than to expose it to the disgusting and disappointing consequences of an attempted marriage."[65] Lawson Tait, a gynecological surgeon and leading British authority on hermaphroditism often cited in American journals, agreed. He encouraged parents to raise children of doubtful sex as male to avoid "lamentable mistakes." By the time a boy reached puberty, he would have learned "whether or not he has marital capacity," and he could avoid marriage. Women, on the other hand, remained largely ignorant about their bodies and "enter[ed] the married state with but a very hazy notion of what its functions are," and so they might marry without realizing their potential incapacity for heterosexual sex.[66]

For older patients too, marriage was considered a primary goal of medical care, especially of surgical intervention. Dr. Frederick Hollick, author of a popular nineteenth-century marriage manual, explained his rationale for surgically altering a patient he believed had been mistakenly raised as male. He wrote about a sixteen-year-old whose parents had presumed him to be male because of his penis, which turned out to be an enlarged clitoris. The person came to Hollick's office complaining of abdominal pain, whereupon Hollick surmised that he was actually a woman and that the pain was built-up menstrual fluid. "It was with the greatest difficulty that I could convince the parents that they had mistaken the sex of their child, whom they insisted in considering a boy," he wrote. "I felt certain, however, that the pains complained of arose from Menstruation, and that the usual flow would be seen if the Vagina was not closed."[67] Hollick made an incision in a membrane covering the vaginal opening, which precipitated the menstrual flow. "The only deformity now existing," Hollick opined, "was the enlarged Clitoris, and this, at the earnest request of the parents, was amputated, till it was no larger than usual." Hollick's case was certainly one of the earliest intersex clitoridectomies on record, and according to him, it was a complete success. He boasted, "She was now perfectly female, and, in a short time, little or no difference could be seen between her and most other young women of the same age."[68]

Doctors like Hollick no doubt believed they were making good decisions for their patients. Hollick observed, "If this had not been done, she would always have been considered an imperfect male, or an hermaphrodite, and would have led a life of misery in consequence." He justified his procedure by mentioning, "I have since heard that she afterward married and became a mother." Not only did she begin to menstruate, but ironically, this person's entire presentation changed after the clitoris, typically a sign of womanhood, was mostly removed. Before the surgery, she had short hair, her voice was "rough," and her pelvis narrow. Hollick reemphasized that after the operation, "especially after Menstruation had begun, the appearance changed rapidly, so that in a short time she differed but little from other young persons of her sex. The hair grew long, the voice softened in its tone, and the pelvis rapidly attained its full dimensions."[69]

By the second half of the nineteenth century, an irony had emerged in the medical reporting of so-called hermaphrodites. Most accounts published in leading medical journals argued that hermaphrodites did not exist in the human species and that all such cases were simply patients

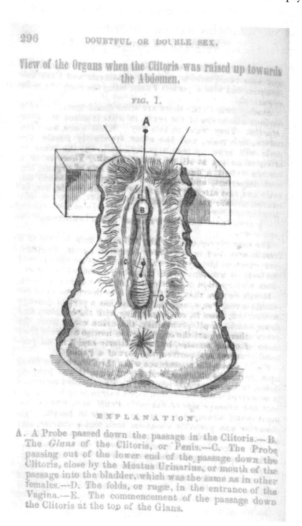

296   DOUBTFUL OR DOUBLE SEX.

View of the Organs when the Clitoris was raised up towards
the Abdomen.

FIG. 1.

EXPLANATION.

A. A Probe passed down the passage in the Clitoris.— B. The *Glans* of the Clitoris, or Penis.—C. The Probe passing out of the lower end of the passage down the Clitoris, close by the Meatus Urinarius, or mouth of the passage into the bladder, which was the same as in other females.—D. The folds, or rugæ, in the entrance of the Vagina.—E. The commencement of the passage down the Clitoris at the top of the Glans.

(*Above and opposite*) Dr. Hollick believed that sometimes the clitoris could grow so large, as in these pictures, that "it could be used like the male organ, with another female, and thus an imperfect connection could be held, but it, of course, could not lead to conception, owing to their being no secretion of semen." Hollick amputated the clitoris of his sixteen-year-old patient (a person whose sex, male or female, had been mistaken. Yet, incongruously, medical men clung to the term, further refining it to justify their pronouncements of one sex or the other. The classification schemes devised by European physicians became more detailed and ultimately, as Alice Dreger has

View of the Organs with the Clitoris hanging down in its natural position, when not Erect.

FIG. 2.

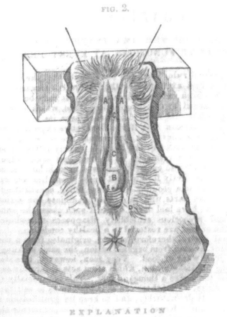

EXPLANATION

A. A. The Large Lips.—B. The Glans of the Clitoris.—
C. C. The body of the Clitoris, or Penis.—D. The Vagina.—E. The opening in the Glans.

It is probable that the Urine actually passed down the passage in the Clitoris when that hung down, but that it passed out of the natural opening, (at C. Fig. 1.) when the Clitoris was held up. There seems little doubt of this organ having been fully capable of the usual functions of the Penis, with another female

raised male but whom Hollick believed female) and recorded that she became more womanly in many respects. Image from Frederick Hollick, *The Marriage Guide; or, Natural History of Generation: A Private Instructor for Married Persons and Those About to Marry* (New York: American News Company, 1860). Courtesy Cornell University Library.

argued, centered on the gonads and required testicular and ovarian tissue to label someone a true hermaphrodite. Since many believed such precise dualism was impossible, doctors had to figure out what combination of organs (along with what conduct) determined true sex. Despite surprisingly little medical agreement on exact criteria, it became the physicians' prerogative to proclaim sex, even if their assessments contradicted how their patients had lived their lives.[70]

Whose Decision?

Doctors recognized that their verdicts had profound significance. Indeed, Theodric Romeyn Beck and John B. Beck asserted, "The decision may be important in deciding the employment in life of an individual, the descent of property, and the judicial decisions concerning impotence or sterility."[71] To facilitate such judgments, Dr. E. Noeggerath commented in the *American Journal of Obstetrics* in 1880, "The testimony of competent medical authority is very essential for the correct and intelligent solution of the legal question at issue." Thus, with the best of intentions, doctors described the facts of their patients' lives—analyzing their physical bodies, personalities, and behaviors—to prove that they were decidedly *not* hermaphrodites and were really one sex or the other, despite ambiguous or mixed genitalia.[72]

In 1863, for example, during the Civil War, a Dr. B. Cloak examined an injured twenty-one-year-old soldier with the intention of returning him to duty.[73] When he discovered the man's indefinite sex organs, Dr. Cloak took on the case for further evaluation. The patient, M. B. H., had lived as a man, though his sexual performance as a male was severely limited. He told the doctor that he "may have had something like an erection once or twice." He had no sexual desire but had once had a nocturnal emission. He had little or no beard but shaved anyway. Ever since he was fifteen years old, he had suffered from what seemed to be a bloody discharge each month, accompanied by back pain, dizziness, and discomfort in his groin. As significant, according to Cloak, M. B. H. could sing soprano as well as bass, enjoyed the company of women more than that of men, had rather full breasts for a man, and exhibited "nearly an equal blending of the male and female natures."[74]

Despite the blending of natures, Cloak ultimately concluded that

M. B. H. was not a hermaphrodite but had a "preponderance of woman." Why did Dr. Cloak feel compelled to make that determination, defining what was not only ambiguous but contrary to M. B. H.'s performative life as a man? Cloak may have been looking for sex organs that were perfect and complete; since M. B. H.'s were not, the doctor could declare that he was not a hermaphrodite, but he then had to evaluate other biological and social indicators to render a decision as to M. B. H.'s sex.[75]

According to Dr. Cloak, the patient was mistaken in living as a man and fighting in the Civil War. The doctor implied that M. B. H. ignored his "preponderance of woman" and that he should have lived as female. He bled every month, could sing soprano, enjoyed female companionship, and had no sexual desire: what better indicators of womanhood? When doctors insisted that their patients had lived lives as the wrong sex, they often implied that the mistake had been intentional. Dr. Cloak wrote of M. B. H. as if he were deceitful: M. B. H. "has always passed" as a man, he reported. Even Cloak's discussion of his patient's expressed sexuality read as if the doctor doubted M. B. H.'s every word: "The statement, *if true,* that he has no sexual desire, is another evidence" of M. B. H.'s womanhood.[76]

Perhaps M. B. H. was unaware that his penis had no urethra and that the curious opening he exhibited was, according to Dr. Cloak, "no doubt a rudimentary vagina," the bloody discharges surely menstruation. The doctor speculated that if M. B. H. would submit to exploratory surgery, doctors would even find "a womb in its proper position." M. B. H. was considered to be a woman in part because "she" had little sexual desire.[77] A patient's sexual desire (or lack of it) was a factor in the medical determination of true sex, and as the century progressed, doctors increasingly considered it, particularly if they were contemplating surgery.

Alice Dreger has termed the years 1871–1915 the "age of gonads" in France and England.[78] By the late nineteenth century, European doctors argued that "true hermaphrodites" were those whose bodies (examined during autopsies) contained both ovarian and testicular tissue. All others, despite unusual conformations of external genitalia, were labeled as either mostly female or mostly male (male pseudohermaphrodites or female pseudohermaphrodites), and hence the two-sex system could remain largely intact. I have argued that in the United States too there was the impetus to maintain a two-sex system, though the impulse began even earlier, and the system used different criteria to establish maleness or femaleness. Before the development of technology capable of analyzing ovarian and testicular tissue, doctors focused on visual markers, particularly the penis and clitoris, though sometimes the vagina, uterus, and menstruation were offered as proof of womanhood. When biological cues proved inconclusive, medical men turned to social indicators—such as a person's mannerisms, clothing, or tastes—to make their determination of sex definitive.

Whatever the markers involved, doctors first debated whether or not their patients were hermaphrodites and then assumed the responsibility of definitively sexing them. Their decisions involved several factors, some medical, some social, and some linked to larger cultural anxieties about racial identity and fraud. American doctors discussed their European colleagues' case histories, read European medical textbooks, and used them to formulate their own conclusions about the impossibility of hermaphrodites. In this country the tendency to proclaim hermaphroditism "impossible" began long before the age of gonads, though it became even more pronounced by the late nineteenth century, when the gonads became crucially important in the United States as in Europe.

CHAPTER THREE

The Conflation of Hermaphrodites and Sexual Perverts at the Turn of the Century

From the anatomical point of view, the hermaphrodite is neither a monstrosity nor a freak of nature, but a creature devoid of ordinary development, that is, not developed sexually in conformity to its species. Physiologically, the hermaphrodite is a degenerate, impotent and sterile, imperfect in impulse and characteristic equilibrium, on account of unstable and perverted sex.

IN THE LAST DECADES of the nineteenth century, the controversy over whether hermaphrodites existed intensified. Doctors on both sides of the issue wrote of actual cases and tried to persuade colleagues of their interpretations of patients' conditions. Despite their fundamental disagreement, doctors on opposite sides of the debate had much in common. Most of them asserted their own crucial role in deciding which gender a patient was. Most of them wanted to see their patients involved in heterosexual relationships, especially marriage. And most of them associated hermaphroditism (to the extent that they admitted its existence) with sexual perversion.

The early examples of interventionist surgery, those undertaken in the late nineteenth and first half of the twentieth century, were designed to serve patients' sexual needs as the doctors in attendance perceived them. Ensuring heterosexual sex, particularly intercourse, grew in importance to doctors. By the turn of the twentieth century, individuals with ambiguous genital conformation were considered potential homosexuals or "inverts"; if people looked both male and female, in their confusion, they might be attracted to people of the "wrong" sex. If an indeterminately sexed patient, even one who seemed predominantly male, expressed sexual interest in men, for example, doctors advocated surgical intervention to make that person's genitals conventionally female in appearance and function. As the twentieth century progressed, medical authorities became more and more certain of their ability to determine a person's sex, and with improved surgical techniques, they believed they were increasingly able to impose a genital conformation that suited their prejudices against same-sex unions.

To uncover the history of intersex is to expose the consequences of pervasive attitudes toward permissible and impermissible sexuality. What doctors began in earnest in the nineteenth century—the surgical "correction" of genitalia to match entrenched notions of normal bodies—became far more routine and troubling in the twentieth century.[1] Physicians uncritically accepted and acted on their own heterosexual norms, often elevating marriage as a goal to parents considering surgery for their children.[2] Heidi Walcutt, in an autobiographical essay, recalled that in the 1970s a doctor at the Buffalo Children's Hospital told her at age fourteen or fifteen that she would need surgery to increase the depth of her vagina "if you ever want to have normal sex with your husband."[3] The impulse to ensure that women would be penetrable began long ago, in the nineteenth century. Promoting marriage was one important objective; avoiding homosexuality was another.[4]

In this chapter we will see what criteria late-nineteenth- and early-twentieth-century doctors began to use to distinguish between alleged hermaphrodites and homosexuals. To modern readers, the distinctions seem obvious: hermaphroditism concerns physical anomalies. People born with ambiguous genitalia may or may not experience same-sex desire, just as people born with unequivocal genitalia do. Homosexuality has nothing to do with the shape of one's genitals; it concerns sexual orientation. Though scientists today search for genetic or hormonal clues to homosexuality's origin, thus connecting propensity to biology, the predominant view is that same-sex desire is not dependent on bodily structure. Scientists investigating biological causes of homosexuality do not typically study genital anatomy.[5]

In the late nineteenth and early twentieth centuries, some doctors drew a distinction similar to the one commonly made today: hermaphrodites by definition had unusual genital anatomies; homosexuals could, but did not necessarily, have unusual genital anatomies. The divergence, however, was not clear-cut. Doctors often equated hermaphrodites and homosexuals. For example, they might ask: If hermaphrodites' genitals were ambiguous, masking their true sex, did their sexual intercourse constitute homosexuality? And since homosexuals preferred intimacy with members of their same sex, then might they suffer from a "mental" or "psychical" hermaphroditism, a deformity centered in the brain, which is also an organ? Though possessed of normal genitalia, such deviants (as they were then called) might nonetheless be pushed by hermaphroditic mental organs to feel and act transgressively. Adherents to this opinion wanted to amend negative views about same-sex desire by proving it was congenital—just as many today assert that sexual proclivity is biological, innate, and immutable rather than "a choice," which is subject to condemnation as immoral. Where, when, and how did human mind and matter interact to produce acceptable or deviant sexual behavior? they asked. Other physicians had no such agenda, and they may have believed that homosexuality was a freely chosen vile outrage. In the conflating of categories, hermaphroditism became further entangled with negative associations of degeneracy. Whereas hermaphroditism was earlier connected with monstrosity and duplicity, in the late nineteenth and early twentieth centuries, it became virtually synonymous with immorality and perversion. Doctors stumbled through these questions as they encountered a real world more complicated and various than the neatly binary one they preferred and prescribed.

Did Hermaphrodites Exist?

In 1879, the president of the Obstetrical Society of Cincinnati, Dr. J. W. Underhill, posed the question: "Is there such a phenomenon as a hermaphrodite among the human species?" He thought there was, but he admitted that the question was controversial and confessed that there was no easy answer. The answer depended on the definition of hermaphrodite. He acknowledged earlier authorities who believed that a hermaphrodite must not only have the genitalia of both sexes, but "must be capable of both begetting and conceiving." That, he agreed, was impossible in humans, as only creatures lower on the animal scale such as the zoophytes and mollusks were capable of a dual reproductive role. But, if hermaphrodites were defined more broadly, he granted their existence: "An hermaphrodite is an animal in which there exists *a mixture* of the male and female organs." A mixture of sexual organs, not the complete capabilities of both sexes, demarcated the hermaphrodite.[6]

Underhill went one step further in expanding criteria. He divided hermaphrodites into four classes: those with dominant male organs; those with dominant female organs; those he called neuters who seemed to have no sexual organs at all; and those for whom the mixture was such that the sex could not be determined, either during life or via postmortem examination. Underhill's definition was broad in that it would have included many people with unusual genital anatomy. But reading closely, it is clear that his aim was not inclusion but taxonomy. As Underhill stated with regard to his last category, classification was essential, even though "it may be impossible to know, either during life or upon post-mortem, to which sex the being *more properly belongs.*"[7] The sense that each individual "belonged," or "properly" should belong, to the male or female category was pervasive.

We can see how doctors and laypeople would assume the existence of two distinct sexual categories. Most people were readily sorted into one group or the other, with no apparent doubt as to their classification. But a rigid binary division had consequences for those whose sex could not be easily determined. Could one legitimately remain outside this inflexible bifurcation? If not, and one had to be designated female or male, how would that choice be made? Would doctors choose for their patients? How would doctors deal with patients who, they believed, were living as the wrong gender? Would they persuade patients to switch genders in midlife? Would they urge surgery to make their patients' bodies fit more neatly into one or the other category? And how would people who were accustomed to their genitals respond? Would they listen to their doctors? Or would they ignore their medical advice and continue living their lives as they chose? What would be the criteria for choosing a gender? Would doctors and patients base their decisions on secondary sex characteristics, on gender performance, on sexual desire?

Those authorities who believed in the impossibility of hermaphrodites did not deny that some people were born with a mixture of male and female organs. Rather, like Underhill, they held that, despite such bodily conformations, each person had a true sex to which he or she belonged. Doctors must look harder to determine each such patient's core identity. According to Dr. J. W. Long, professor of gynecology in Richmond, Virginia, in an 1896 article aptly titled, "Hermaphrodism, So-Called," there were two types of hermaphrodites: true and false. True hermaphrodites, people with both an ovary and a testicle, were "so rare in the human species as to be almost a non-entity."[8] False hermaphrodites, sometimes called spurious or pseudohermaphrodites, were more common; either they were truly males who resembled females or truly females who resembled males.[9] All agreed that, except for the all-but-nonexistent true hermaphrodites, a person's real sex could be uncovered and that if those with ambiguous or nonconforming bodies were left alone, not only would they suffer unhappy, unfulfilled lives, but all manner of unruly behavior, particularly deviant sexual conduct, would probably ensue.

Links to Homosexuality

If hermaphrodites were exceedingly rare, then ambiguously sexed patients were simply men or women who needed to learn their true sex.[10] Of course, that knowledge could lead to trouble. Suppose, for example, patient X was living as a male in sexual intimacy with a woman. If a doctor decided patient X was not a man, but a woman, X's relationship was homosexual. Hermaphroditism thus might foster homosexuality and, since homosexuality was a prohibited "vice," perversity and immorality.

The Chicago urologist and surgeon Dr. G. Frank Lydston shared the popular view that true hermaphroditism—the presence of perfect sets of male and female genitals in an individual able to procreate both as a man and as a woman—could not exist. And, since he believed the "so-called hermaphrodite" was sterile, it would be better to think of the patient as *neither* male nor female, rather than *both* male and female.[11] Some, he argued, did not have the "desire or capacity" to have intercourse with anyone of either sex. Others would be able to use their ambiguous genitals to perform sexually *as* men *and* women, *with* men *or* women, despite their inability to procreate. Pseudohermaphrodites (a term generally used to refer to individuals with the internal organs of one sex and the external genitalia of the other but often used more loosely) might succumb to sexual perversion through "an apparent commingling of the functional capacity." In other words, some so-called pseudohermaphrodites were capable of having sex with both sexes because their genitals could function both penetratively and receptively.

Lydston discussed what seemed to be such a case. The subject was a mulatto cook[12] with hypospadias, a condition in which the urethral opening is not at the tip of the penis, but underneath. In severe cases, the opening falls close to the scrotum, which can resemble the labia majora.[13] The cook had been accused of infecting young boys in the neighborhood with gonorrhea. Some investigation found that he had contracted gonorrhea "in the normal manner," from a woman. Then, "performing the passive role" with the boys, he passed it along to them. The doctor suggested a "commingling" of function. The cook could use his hypospadic penis to penetrate women "normally" for heterosexual sex, and perhaps a vaginal opening, indicating another intersex condition, allowed him to be a receptor in homosexual encounters.[14]

Many physicians agreed that alleged hermaphrodites were most commonly women with enlarged clitorises. Dr. George DuBois Parmly's definition of the phenomenon echoed the eighteenth-century model: a hermaphrodite was simply "a person of female sex possessed of a clitoris sufficiently developed for a sort of spurious coitus." Parmly believed that if a woman so constituted could pass as a man, "she" would. Given women's subordinate social roles in the late nineteenth century, it should not be surprising that doctors assumed that if given the choice, hermaphrodites would rather live as men. In 1886, Parmly reported in a major obstetrics journal about "a perfect hermaphrodite, in so far, at least, as the power of performing the sexual act, for s/he had lived with men as mistress, and with women as lover. Being asked which s/he preferred, to be a man or to be a woman, s/he answered: 'To be a man, for it gives greater social independence.'"[15] The desire for social freedom, Parmly postulated, compelled some "strong-minded" women to "masquerade as men." Other women were motivated in their charade by the fear of childbirth, he suggested. Focusing on genital anomalies that would make heterosexual sex uncomfortable, Parmly suggested that sex with women (he called it Sapphism) could be more pleasurable for hermaphrodites than sex with men: "Perhaps, in some cases, the presence of an elongated and voluminous clitoris which renders sexual congress painful or impossible, while Sapphism is practicable and, in a degree, pleasurable, may decide a woman to take the role of a man in social life."[16] For Parmly, social and sexual roles merged, and hermaphrodites became Sapphists. Though he was one of the rare doctors who seemed to sympathize with a subject's desire for sexual pleasure, Parmly did not countenance same-sex relationships.[17]

Expanding on his discussion, Parmly explained that hermaphrodites sometimes became "the lovers of women, or, what is rarer, [took] up the role of woman with some male lover." He recounted a newspaper story of a young woman who had abandoned her husband and child to become the "husband" of a young lady: "They had gone before a clergyman who suspected nothing, married them, making them 'man' and 'wife,' as far as lay in his power . . . They went at once to housekeeping, and seemed quite pleased with each other, when the real husband came along and electrified the community by announcing that this 'husband' was his own wife and the mother of his child. It must, indeed, be a *peculiar moral condition* which renders such *ill-assorted unions* possible."[18] Without mentioning any particular medical condition the person in the news account might have had, Parmly's editorial comments simultaneously condemned homosexuality and linked hermaphroditism with family desertion and "peculiar" morality.

Doctors depicted hermaphrodites and homosexuals alike as exemplars of degeneration—biological as well as social harbingers of family destruction. Parmly retold several such stories that, he admitted, "could not be vouched for." One began with an ordinary couple who had children and lived happily together. The supposed husband, as it turned out, "was not as other men." Apparently, he menstruated once a month and was called a hermaphrodite. Parmly concluded, "Even accepting the presence of a long clitoris, we have drawn the picture of a woman, and not of an hermaphrodite. Of course, the impregnation of the wife was quite possible, but we must look for the father of the children—outside. This is another variety of spurious hermaphrodites, in fact, merely women disguised. I think it very probable that a large and long clitoris existed." Despite the lack of physical evidence and firsthand knowledge of the events, Parmly confidently outlined the associations between same-sex relations and hermaphroditism.[19]

The unification of hermaphroditism with homosexuality was more easily accomplished because Parmly and other doctors did not always use genuine cases to make their points. Physicians published anecdotes they had heard or read about, not limiting their analyses to patients whom they had examined. Stories like the ones above proliferated in nineteenth-century medical literature. Filled with speculation and innuendo, the image of the hermaphrodite as a deceiver persisted. The narratives confirmed what most readers probably intuitively believed: that since true hermaphrodites could not exist, those supposed hermaphrodites who lived in same-sex unions were "passing" and trying to get away with illicit perversion.

In the late nineteenth and early twentieth centuries, several authors went beyond reporting undocumented anecdotes to make the link between homosexuality and hermaphroditism scholarly, medical, and presumably indisputable. This association turned on the nature and origin of homosexuality: was it a congenital defect or an acquired condition? Medical theorists debated this question. Havelock Ellis, the British sexologist whose six-volume work *Studies in the Psychology of Sex* was published in 1897–1910 and widely read by American theorists, advocated what was at the time a progressive position on homosexuality, emphasizing its congenital nature. But the language he used easily slipped into a discourse of pathology. To explain how a congenital predisposition could evolve into homosexuality, Ellis invoked unfavorable conditions, such as same-sex environments, criminality, weak will, or other degeneracy. According to the historian Jennifer Terry, Ellis wanted it both ways: he argued against associating homosexuality with immorality but ultimately did just that.[20]

Those writers who believed homosexuality was congenital often conflated it with hermaphroditism. Lydston was one such writer, an influential medical doctor who believed that sexual perversity (which included, but was not limited to, homosexuality) had to be tied to either brain or genital conformation. Lydston positioned himself as a man of science, eager to attribute what he saw as perversity not to immorality—to a "willful viciousness over which they [homosexuals] have, or ought to have volitional control"—but to physical causes that, potentially, could be amended. Lydston suggested two sources of sexual perversion: congenital and acquired. He linked the former to biological defects and suggested that congenital sexual perversion could manifest itself in a "defect of genital structure, e.g. hermaphroditism."[21]

Lydston went one step further, suggesting that just as physical hermaphroditism was caused by improper genital differentiation, certain cases of "sexual perversion" could be attributed to a similar "imperfect" separation in the brain. In other words, whereas homosexuals' genitals might look "normal," something in their brains might have developed improperly. Worried that his theory might seem a "trifle far-fetched," Lydston explained that even when the physical demarcations of sex are complete, the "receptive and generative centers of sexual sensibility may fail to become perfectly differentiated." The result could be either sexual apathy or same-sex desire. Such congenital lack of differentiation, he suggested, would be impossible to detect physically, but it could be responsible "for disgusting cases of sexual perversion that society is prone to attribute to moral depravity."[22]

Thus doctors could at least look to a suspected homosexual's genitals to see if they were the cause of the subject's homosexuality. If patients were homosexual and yet had normal-appearing genitalia, then Lydston suggested that their brains, structurally hermaphroditic, had an imperfect separation. Similarly, doctors might suspect that an infant's atypical genital structure could lead to homosexuality at sexual maturation. Because of the emphasis on the congenital nature of both hermaphroditism and homosexuality, the terms became nearly synonymous, and people characterized by either were sometimes called "inverts."

By the 1890s, fears of homosexuality intersected with anxieties about hermaphroditism, the anatomical confusion of the latter merging with the supposed social and sexual confusion of the former.[23] Consider the case of Viola Estella Angell. According to Dr. C. W. Allen of New York, who published the report in 1897, Angell dressed and looked female, and being destitute, she applied for admission at the "Florence Mission, a shelter for fallen women."[24] But Angell was no ordinary woman, and after a physical examination at the institution, she was rejected and sent to a sanitarium, where Dr. Allen examined her. She had been raised a girl, but at puberty her genitals underwent "certain changes" that prompted her mother to dress her in boys' clothes. According to the report, the mother thought the child "would have less trouble in the world as a male." Unfortunately, that was not the case. When he was sent off to school, other students constantly called him "sissy" because of his "effeminate ways, manners, and general appearance." He led "the life of a dog from the taunts and jeers of his school-mates" and left school with "thousands" running after him, accusing him of being a girl. Was he a girl in boys' clothing?[25]

Angell had the physical characteristics of those labeled "hermaphrodite." According to the physician, "he" had a penis and bled monthly for several days each month, the blood issuing from the rectum and the "penile urethra." Though she had been declared "unsuitable" for the mission because of her physical conformation, Dr. Allen said that she entered the sanitarium willingly, "seemingly desirous of having the question of sex definitely settled."[26] As for sexual relations, Dr. Allen reported that "the instincts are in every way those of a woman, and it is denied that any manly feelings exist." By this the doctor meant that not only was Angell attracted to men rather than women, but also, consonant with impressions of diminished female sexuality current in the Victorian period, that Angell was deficient in sex drive altogether. Though Angell had engaged in sexual intercourse with men, the doctor noted, it was "always rather against his inclination."[27]

Angell was an enigma to Dr. Allen. With no manly feelings (indeed, Dr. Allen reported that Angell's penis had never been erect) and a "repugnance for man's occupations," "he" was not typically male. It was no surprise that "he" would run away from home dressed as a woman. But neither was Allen entirely convinced that Angell was a woman. What indicators did he observe? Physically, while one hand and foot looked feminine, the other side looked more masculine; the pelvis and trunk, too, were masculine in type. His face (Allen referred to Angell with masculine pronouns) was "distinctly feminine," his voice soprano, and his gait, manner of sitting, standing, and sewing were "essentially feminine"— characteristics Dr. Allen believed were innate rather than acquired. Dr. Allen described the conventional qualities of a woman, but since he regarded Angell to be really male, his assessment can be read as a recitation of the stereotypical traits of a homosexual man. For example, Angell showed a "fondness for personal adornment with gay colors," as well as for "neat and clean bows, ruffles, and ribbons, finery, and flowers," and he was "slightly hysterical at times." He liked poetry, music, acting, drawing, and writing letters, especially on topics that required expressiveness, such as love.[28]

Dr. Allen tried to make sense of Angell's condition, which, given the patient's womanly emotional characteristics and sexual inclinations and a bodily conformation that included a penis, seemed to go beyond simple genital confusion. He asserted that Angell had a "pronounced state of *mental* hermaphroditism."[29] Angell sincerely believed she was a woman. She was not using a ruse in order to have sexual intercourse with men, as Allen believed Angell had little if any sex drive. Simply put, the patient had "a desire to engage in a legitimate law-abiding manner in the pursuits of the sex whose instincts the subject feels." These instincts were innate, something Angell was born with; even though her genitals suggested she was "really" a man, she had the social proclivities of a woman.

Dr. Allen's assessment sounds strikingly similar to contemporary descriptions of transgender identity: Angell was "honestly convinced that nature had intended him for a female." As the historian Joanne Meyerowitz has pointed out, early sexologists considered cross-gender identification to be a category of "inversion"; it wasn't until the mid-twentieth century that "transsexualism," the term for sex change through hormones and surgery, was coined.[30] And so, confronted with the patient's aberrant physical and mental presentation, Dr. C. W. Allen applied a relatively new term, "psychical hermaphroditism," or "mental hermaphroditism," to describe men who believed they were meant to be women.

One month earlier than Dr. Allen's report, Dr. William Lee Howard had published an article called "Psychical Hermaphroditism," in which he distinguished inversion from the possession of unusual genitalia, but he nonetheless applied the term "hermaphroditism" to what he saw as "perversion." According to Howard, the genitals of the "pervert" are typically "normal in appearance and function."[31] Inversion, then, was purely a psychological condition and not something structural. Like hermaphroditism, though, inversion was congenital, rather than acquired. Howard quoted one of his patients, a thirty-year-old man, who said he loved men just as other men loved women. The patient explained, "I can define my disposition no better than to say that I seem to be a female in a perfectly formed male body, for, so far as I know, I am a well-formed man, capable of performing all of man's functions sexually. Yet as far back as I can remember, surely as young as nine years, I seemed to have the strongest possible desire to be a girl, and used to wonder if by some peculiar magic I might not be transformed. I played with dolls; girls were my companions; their tastes were my tastes; flowers and millinery interested me and do now."[32] Thus a new twist on hermaphroditism emerged, an inversion that involved the mind rather than strictly the body.[33]

By the end of the nineteenth century, "psychosexual," "mental," or "psychical" hermaphroditism were all terms doctors used to describe patients who admitted to same-sex desire. People with normal genitalia who confessed to an "inverted" gender identity, like Dr. Howard's patient, were termed "hermaphrodites," as were people with ambiguous genitalia, whatever their sexual inclinations. Hermaphroditism became a term that could be used to describe either a physical condition (with or without homosexuality) or a psychological one involving same-sex desire.

For the most part, doctors considered physical hermaphroditism to be a congenital structural malformation, apparent at birth. However, some physicians encountered patients who claimed that their bodily organs were typical at birth yet changed shape during puberty. Angell's mother, in Dr. C. W. Allen's case, had said that Angell's genitals underwent "certain changes" at puberty, which prompted the mother to start dressing her in boys' clothes. Some doctors described how patients' bodies seemed to shift suddenly from female to male, for example, acquiring male primary and secondary sex characteristics.[34] If a person with 5alpha-reductase deficiency (the metabolic anomaly featured in Jeffrey Eugenides' Pulitzer Prize–winning novel, *Middlesex*) was assumed to be female and the body began to alter when testosterone produced virilization, then sexual relationships previously considered heterosexual would become homosexual.

Doctors wrote about these cases of sudden genital diversion without the scientific objectivity that we might expect from medical reports. Instead they emphasized the sexual and romantic repercussions. In the earlier part of the nineteenth century, when race was the central preoccupation, such cases of sex metamorphosis highlighted anxiety about racial transmutation. In the late nineteenth and early twentieth century, when issues of sexuality, licentiousness, and freedom from social restraints had come to the fore, physicians dwelt on the sexual possibilities available to people who found their bodies undergoing radical transformations.

Dr. J. B. Naylor's 1896 account of a case in Ohio is typical of articles stressing the striking nature of the change and the sexual implications that arose from it. "X has had somewhat of a romantic history," Naylor began, "Up to the age of seventeen he wore feminine apparel and bore a feminine appellation. During all these years he was to all appearances, intents, and purposes, a female. He performed the duties of a domestic, learned to sew and knit, took a female part at the rustic plays and dances of the neighborhood and slept with his girl friends. But presto! change! all at once—without previous symptom or warning—he blossomed out as a full-blown male." X now wore men's clothes, began to chew and smoke tobacco, played the fiddle for country dances, and associated with "naughty men and [started] to ape and enjoy their rough ways." As if that were not shocking enough, Naylor exclaimed, "at the age of 35 he married—a woman!" After three months, X's wife sought a divorce on the grounds that her husband was impotent. She testified, according to Naylor, that "owing to his unsatisfied sexual desire, he would not let her sleep at all."[35]

X was forty-five years old when Naylor first examined him. "His form from the waist upward is markedly masculine, and from the waist downward more or less feminine. The legs are short, the hips broad, and the thighs plump and tapering. His feet are small, and he walks with the mincing steps of a female," Naylor wrote. X perplexed Naylor, as his assessment of him as *both* and *neither* male or female attested. "So far as the external organs of generation are concerned, he is both male and female, but so far as the act of generation is concerned, he is neither." Naylor explained that though X had a penis, testicular tissue, and "male passions," he could not ejaculate because his penis was imperforate. In addition, he had female external organs, "suffers from headache and backache once a month, voids his urine as a female, and walks and talks like a woman." These characteristics led Naylor to believe that he also had a "rudimentary uterus and ovaries."[36] Yet Naylor's use of male pronouns rendered X male, as did the story's dramatic emphasis on the shift from female to male, a change resulting in a heterosexual marriage, short-lived and frustrating though it may have been.

Exposing these astonishing transformative conditions enhanced the conflation between hermaphroditism and deceptive sexual perversion. One anonymously published article with the pejorative title, "Hermaphroditism, or Sexual Perversion," described how allegations of hermaphroditism may have been merely covers for homosexual liaisons. In this 1890 account, a twenty-two-year old man had suddenly "blossomed forth" as a woman, assumed the suggestive name Belle Hardman, and married a "respectable blacksmith." Hardman's mother and husband were prepared to testify that he had indeed been born and raised a boy and had lived as a male for twenty years until his body underwent this unusual transformation into womanhood. The article claimed that even a San Francisco doctor "vouches for the marvelous change of sex as an absolute fact."[37]

Could Belle Hardman have been afflicted with an intersex condition that manifested itself during puberty, when hormonal factors may alter the outward physical shape of a person's body? Or, was the alleged hermaphroditism just a cover for a same-sex sexual relationship? At least one doctor involved in this case seemed to think that fraud was involved. The family physician, Dr. Henry A. DuBois, said that he never noticed any malformation in the boy as a child, though he had never examined his genitals. Dr. DuBois believed "Belle" was all male and suggested that the "present association is due to unsoundness of mind" or perhaps a publicity stunt. According to the article's anonymous author, only time would resolve this puzzle: would Belle Hardman get pregnant and become a mother? In the meantime, the author admitted perplexity: "The doctors, the neighbors and all the country round are asking, 'What is it?'"[38] Stories like these, often written anonymously though appearing in respectable medical journals, conflated hermaphroditism with homosexuality in physicians' minds.

Surgery, Celibacy, and Heterosexuality

Doctors believed that surgery was warranted in many cases of atypical genitalia, not necessarily for the health, comfort, or pleasure of the patient, but to preclude the undesirable potential for homosexual sex. Even lifelong celibacy was preferable to homosexuality. Physicians in the 1880s and 1890s wanted their patients to understand their hermaphroditic conditions as deformities and not as a physical license to commit sexual immorality. Dr. J. W. Long wrote that "the peculiarities which make them appear mixed, are only deformities like hair-lip or club-foot . . . I believe that we owe it to these poor unfortunates to impress upon them, as well as upon others, that they are NOT part man and part woman; but that they always, with scarce an exception, belong to either one sex or another."[39] Doctors agreed that whatever sex their patients "truly" belonged to, they should choose a partner of the opposite one for sexual gratification. If this desirable pairing proved impossible, some physicians advised castration to eliminate "homosexual" desire.

A case Dr. Samuel E. Woody of the Kentucky School of Medicine presented in 1896 exemplifies this inclination to use surgical treatment to purge same-sex proclivity. A young woman, aged twenty, came to see Woody in his office. She introduced herself as a hermaphrodite and offered to submit to an examination "in the interest of science and her own purse." Perhaps she wanted to exhibit herself as an hermaphrodite, a common freak show spectacle at carnivals and circuses, and sought a doctor's letter attesting to her condition, providing credibility to her performance.[40] Skeptical of her motives from the outset, Woody said that her statement that she had been reared a girl, "of course, must be taken *cum grano salis*." According to her narrative at puberty a penis appeared, and, indeed, Woody found a mixture of genitalia, including labia and a testicle. The clitoris "stretched itself out into a penis," and Woody characterized its erection: "Rising up thus and stooping forward with its great mantle of loose skin, the effect was ghostly."[41]

Woody's suspicion about this woman persisted, and ultimately he became convinced that "the young *lady* was a young *man*," probably because of the testicle and lack of ovaries, despite her outward feminine appearance and female gender identification. She had had sex with men and women and said she was equally satisfied sexually with both, though lately she preferred sex with women. If she were a woman, as she presented herself, she was enjoying homosexual relationships with other women. If, as Woody believed, she was "really" a man, then "he" also had had successful sex with other men. Either way, Woody was not pleased. And so, in principle, he recommended eliminating sexual desire altogether through surgery: "So ill-fitted for the generative function and so prone to psychical perversions and moral degradation, such cases should be castrated in early life."[42]

Some doctors were equivocal about treatment options when confronted with unusual genitalia, but many shared the conviction that "unsexing" a patient was an appropriate measure to prevent what they saw as impending immorality. An article in the *Medical and Surgical Reporter* recounted a London case in 1884 in which the doctors considered their options regarding an otherwise healthy nine-year-old girl. Though the child looked like a little boy when fully clothed, closer examination revealed that she had female external organs. Internally, the doctors felt what they thought was a testicle, and they wanted to remove it to be sure. They justified the procedure on moral grounds. If the body they removed turned out to be an ovary, then their actions would have helped the girl avoid painful menstruation when she matured sexually. If it turned out to be a testicle, they argued, their procedure was even more necessary because then the child would have been considered a boy. Upon sexual maturation, being a boy with female external organs would have "most complicated results." And so they reasoned, "*morally as well as physically* it was prudent to unsex the person, so far as that can be done by removing the most char-acteristic sexual organs." The organ turned out to be a testicle. That result made the doctors "quite convinced of the propriety" of removing the other testicle as well in another operation a few weeks later.[43]

What is noteworthy here is not that the doctors removed a child's tes-ticles, but that the justification they used for the procedure rested on hypothetical scenarios regarding her (or his, they feared) sexual future. This case is very much like the one Samuel Gross reported in 1852, in which Gross removed a toddler's testicles to insure her future marriage-ability. By the 1880s, the explicit promotion of marriage had shifted to an implied, and often unequivocal, avoidance of homosexuality. In the con-text of surgery and atypical bodies, discussions of "morality" meant keep-ing sex heterosexual.[44]

From the nineteenth century until today, doctors have performed sur-gery to "correct" genital anomalies even when such operations are not required for the patient's voiding. As these examples make clear, surgery has reinforced not only a binary division of the sexes, the creation of perfect males and females, but also traditional gender roles: it has pro-duced women and men who could uphold specific cultural norms. Women should marry, and that destiny implies a need for a vagina that can accom-modate heterosexual intercourse. Men should have penises that allow them to void standing, discharge semen, and penetrate women. Surgery was dictated by societal, rather than simply medical, demands. As these cases illustrate, for some time doctors have been in the business of creat-ing women and men, based on normative models of sex differences and on the elevation of heteronormative marriage.[45]

In the medical discussion of hermaphroditism published in medical journals, we can see how, rather than merely describing atypical conformation, doctors created, or at least reinforced, the notion of the freakish or perverse hermaphrodite. The hermaphrodite could be treated and normalized through surgical procedures, which would culminate not only in satisfactory genitalia, but also in the performance of the suitable social role in marriage. Surgeries typically offered to women included opening of vaginal occlusion, testicular removal, and clitoral excision. Surgeries for men with hypospadias included straightening the penis and extending the urethra through the penis. All of those operations made the patients "fitter" for heterosexual penetration, and the doctors wrote their case studies with intercourse and marriage as the primary indicator of a successful outcome.

One woman in Little Rock, Arkansas, saw several doctors in 1885 after she discovered, upon her marriage at age seventeen, that there was no entrance to her vagina. One doctor told her that she had no womb. Another told her husband that she was "of no sex." Depressed, she sought yet another opinion, and this time the doctor assured her that she was female, as he manually felt her uterus in the normal position. The doctor, E. Cross, advised her that surgery to create a vaginal opening was possible, though he warned her of the operation's possible complications and potential failure. Cross admitted that the procedure was "one which gives the gynecologist as much, if not more trouble and uneasiness than any that falls under his care . . . There are no landmarks to guide him; carefully he must feel his way without compass or chart, and where a slip of the knife or a tear of the parts may at any time cause serious trouble."[46]

As Cross wrote, serious trouble could indeed ensue from invasive gynecological procedures. Patients suffered inflammation and infection from unsanitary conditions and could bleed to death during the operation or after they left the doctor's office.[47] Doctors recognized the dangers and cautioned against certain procedures. In an 1884 book, Dr. Thomas Addis Emmet, owner of a private hospital for women in New York, admitted, "Under the guise of surgery, the uterus has been subjected to a degree of malpractice, which would not be tolerated in any other portion of the body . . . No portion of the body has suffered more," he said, "from the overzealous interference of ignorant practitioners."[48] Any surgery was perilous. Dr. George Tully Vaughan, a founder of the American College of Surgeons, wrote in 1908 that "every one knows something about it and all are doing operations." On the one hand, much good was done and lives were saved. On the other hand, he confessed, "That much harm is done, that much suffering is caused, and that many lives are brought to an untimely end without question is equally true . . . Too often a surgical operation is like a game of chance in which the patient's life is at stake."[49]

Undaunted, the newlywed urged Dr. Cross to proceed. She was willing to take any chance " 'to be made a woman,' as she said." The operation was successful. Using "finger-nail and blunt instrument," Cross was able to create a vaginal tract. Fully loaded with opiate, the woman was given instructions to dilate the newly shaped vagina twice daily. This was apparently a painful procedure, because the doctor noted that one night the woman convinced the nurse to forego the process, and immediately the passage closed up again. A second operation followed, after which Cross put in place a large glass tube, which the woman wore all the time for five weeks. When she left for home, Cross advised her to wear the tube consistently for the next six months. Finally, one year after the second operation, Cross received a letter from the bride reporting that she was in good health, that she menstruated regularly, and, according to Cross, "was in every way a satisfactory wife."[50]

Doctors sought their patients' consent in the late nineteenth and early twentieth centuries, and they often went beyond reporting permission, asserting that their patients insisted upon whatever procedure transpired. That consent was obtained in these cases should not weaken misgivings about the doctors' priorities. We cannot be sure how the patients arrived at their decision for surgical intervention because our sources, including those ostensibly in the patients' voices, derive from physicians. No doubt some people wanted surgery, however dangerous, to accommodate heterosexual relations, as doctors claimed. Yet such decisions were not made in a social vacuum. Negative associations between hermaphroditism and homosexuality probably made ensuring heterosexual norms attractive to patients as well as to their doctors.

In 1884 a young woman, A. B., came to see Dr. William P. McGuire of Winchester, Virginia, to "have the sex to which she belonged determined." Why would she need a doctor's assessment? She had lived as a woman for thirty-five years. The doctor described her feminine features: she was "fairly formed," five feet four inches in height, had no facial hair or

Adam's apple, and had small breasts and long hair cascading down her back. A. B. also had some masculine characteristics: her hands, arms, and legs looked masculine, but most notably, she had a small penis, about three-fourths of an inch long, with no aperture. The urethra was located between the penis and the anus. Upon examination, Dr. McGuire found two testicles, which no doubt confirmed his diagnosis that A. B. was not female at all.[51]

When Dr. McGuire inquired as to A. B.'s sexual proclivities and learned that her "desires had been masculine" (that is, she was attracted to women) and that she sometimes experienced pleasurable sexual sensations at night (alluding to nocturnal emissions), he declared that "there was no trouble in determining her sex." She was, in fact, a man. He had simple words of advice: "to change her dress to that of a man" and to undergo surgery so that a new urethra could be made such that she could urinate in a standing position. As a woman A. B. urinated in the typical sitting position, but as a man that would not do, according to Dr. McGuire, and so he suggested surgery.[52]

According to the contemporary intersex activist Cheryl Chase, since the mid-twentieth century, intersex treatment protocols have been defined by societal expectations of gender performance. The same was true of nineteenth-century treatments. In McGuire's case, the insistence that a man stands rather than sits to urinate is surely a cultural preference and convenience rather than a biological imperative of maleness. Scholars of disability theory, such as Shelley Tremain, have further argued that medical professionals do not merely study and treat alleged impairments, but instead actually create the social category of "impaired."[53] This is not new to the twentieth century. A. B. effectively became impaired upon consultation with a physician. When she arrived at McGuire's office, he remarked that she was "in good circumstances in life." When she left, she had been told to change her dress, her mode of life, and to have a new urethra fashioned so that "he" could stand to urinate. There is no discussion in McGuire's report of how A. B. took the news that she had testicles and ought to change her gender status. She had dressed as a female since birth, "and her business in life [was] that usually followed by that sex." Now she was being told that she was not only a male but a male in need of serious reparative surgery.

A. B.'s case provoked controversy in the medical world. The week after

McGuire's article, the *Maryland Medical Journal* published an editorial critique of McGuire's conclusion, suggesting that he had mistaken ovaries for testicles and had erroneously considered a hypertrophied clitoris to be a penis. The editor took exception to McGuire's initial account of the diagnosis, particularly his seemingly careless attitude toward the identification of sex. In the editor's assessment, the "male element" was decidedly in the minority.[54]

What most upset the editor seemed to be McGuire's claim that the subject's sex could be decided easily. Quoting another medical authority, the journal insisted that determining the sex of problematic individuals was "a matter of great difficulty." Even a factor that seemed obviously male, such as nocturnal emissions, for example, could be misread. Among "highly erotic women," the secretion of fluid from the "vulvo-vaginal and other glands is not an uncommon experience." The editor explained that the "enormously hypertrophied" clitoris may indeed look like a penis and that "the proclivities and desires of women may, under vicious moral influences and bad associations, assume a masculine type."[55] In a conclusion that recalls the earlier association between clitoral size and same-sex relations among women, he suggested that A. B. was really a woman with erotic desires for other women.

But McGuire disagreed, and he defended himself in the *Maryland Medical Journal* the following week. He suggested that there was a family history he could not disclose that would support his conclusions. And he adamantly maintained that there was no vagina, no vulva, no uterus, and no menstrual function. The penis, though small, was well-formed, he claimed. As an added testament to the accuracy of his own interpretation, he added a final paragraph to prove his point: McGuire stated that he had heard from A. B., and that "he" was now living as a man, had married a woman, and "has regular intercourse with ejaculation of semen, which, owing to the position of the external opening of the urethra falls not into the vagina of the woman."[56] Because A. B.'s "penis" had no urethral opening, the semen could not discharge from there. Apparently A.B. had not consented to any surgery but yet had begun to live life as a man. Perhaps A. B. saw no need for the surgery, as even McGuire attested that except for the semen placement, "both parties claim it [intercourse] is perfect."[57]

Dr. McGuire legitimated his account by including the uncorroborated information that A. B. had been living as a man and had married a woman. The tactic of reporting additional information and offering rebuttals allowed doctors to "prove" that they had made the proper medical decisions. Since the middle of the nineteenth century, when doctors began to perform surgery that would effectively reassign gender, they have included in their medical accounts hearsay information about their patients' "success." Unfortunately, since material from patients themselves is so woefully scarce, our ability to evaluate the doctors' versions is limited. Often the best we can do is to read between the lines of the doctors' accounts and to assess the challenges they faced from patients and other physicians.

Consent and Challenges

A. B. had taken his doctor's advice and had begun to live as a man, but he apparently did not consent to surgery to move the urethra, which the doctor hoped would allow him to stand during urination and to discharge semen from his penis. Refusing surgery remained an option for patients into the twentieth century, and some *did* choose to live with the gender they were used to, despite whatever new medical information doctors revealed to them. In the late nineteenth century, one thirty-four-year-old who had been living as a woman refused a recommended lengthening of her urethra and correction of what the doctors saw as a curved penis because surgery would necessitate entering the hospital as male. Dr. James Little, a professor of surgery at the University of Vermont and the New York Post-Graduate Medical School, had confessed to the patient that the surgery might not be completely successful. He believed he could straighten the penis but probably could not "increase the length or efficacy" of the urethra. Nonetheless, Little recommended the operation, most likely because he saw the patient as a man and assumed that he would want the curvature fixed. But the patient refused. Dr. Little explained, "He could not, of course, enter a female ward, being a male; nor, on the other hand, could he be put in the male ward still clad in the garments of the other sex, and these he objected to laying aside, as he claimed that he would not like to return to his home, even after an operation, dressed as a man after having passed so many years as a woman."[58]

In part, this person's decision to forego surgery was based on her history of living as female. Other factors influenced her choice, including the limitations of the proposed procedure. Since Dr. Little could not lengthen the urethra, the patient would not have been able to urinate standing up even after surgery. And so, according to Little's account, since she still would have to sit to void, wearing women's clothes would continue to be more convenient. In fact, Little mentioned that the operation was extremely difficult, and "if inflammation sets in, and the corpora cavernosa [within the penis] become involved, their structure is likely to be permanently impaired."[59] In addition, the patient did not have time for the operation and convalescence, as her parents depended on her for support. For this particular person, considerations of practicality, convenience, and the possibility of surgical damage merged with the sheer illogic, for her, of switching genders at age thirty-four.

Yet, throughout his account, Dr. Little referred to the patient as male, although Little knew that she was, and planned to continue, living as female. Patients could sometimes defy doctors' suggestions, but doctors often reassigned them anyway, even if only pronominally.[60] In Dr. Little's view, this patient was merely "passing" and had been doing so for years. The doctor had learned that the family suspected something was amiss years earlier. Though the child was proclaimed a girl at birth and had been reared as a girl, between the ages of twelve and fourteen, the child noticed "that he differed from other girls of his acquaintance, and calling his mother's attention to it, she consulted a physician, who, after making an examination, informed her of the nature of the deformity, and assured her that the child was a male."[61]

The parents continued to raise their child as a girl. According to Little, they were "too ignorant to properly comprehend the difficulty . . . and in consequence made no change in his apparel." In his patronizing attitude toward the parents and patient, Little was typical of nineteenth-century doctors who saw such patients as misguided in not adhering to medical advice. The idea of mistakes would come to have profound implications. Doctors interpreted those with unusual genitals as either "really" male or female and understood patients' own reading of their external genitalia as evidence of their naïve foolishness at best and willful obstinacy or perversion at worst. Even before the early twentieth century, when surgical techniques offered doctors the capacity to create "normal" looking genitals, doctors tried to correct alleged mistakes in patients' self-presentation by social means, suggesting and perhaps pressuring patients to change clothes and genders so that their gender presentation would match the doctors' (rather than the patients') interpretation of their bodies.

Despite their unabashedly certain pronouncements, doctors sometimes conceded that determining a person's sex could be difficult. In 1879, J. W. Underhill, president of the Obstetrical Society of Cincinnati, admitted that "it is not always an easy matter to determine the sex, and mistakes have been made by eminent physicians, some of which errors could not be proven until after the death of the being."[62] Though doctors understood that they could make mistakes, they generally believed that medicine had the answers, and that it was the patients who were mistaken as to their own genders.

In the case of the thirty-four-year-old woman whom Little believed was male, the physician censured both the parents and the patient: "As a result of this stupidity on the part of his parents and his own [the patient's] modesty and want of courage, together with an amount of religious superstition seldom met with, he has grown to his present age, still wearing the garb of his mistaken identity, and passing as a female among his acquaintances; although he is aware that it is generally whispered about the town by many who know him that he is an hermaphrodite."[63] Despite the information offered by the doctor, the patient did not want to change her ways. She might not have enjoyed neighborhood gossip suggesting that she was a hermaphrodite, but there was no guarantee that surgery would eliminate rumors.

Dr. Little's patient was not against all forms of surgery; to the contrary, she wanted surgery that would make her more like other women. According to Little (who still insisted on male pronouns), "His sexual desires were very strong, and were a source of almost constant annoyance to him, as he associated continually with females, but he had never made an attempt at sexual intercourse for fear of exposing his true condition." The patient had initially written to Little asking about the possibility of castration, "with a view to putting an end to his sexual appetite, and he stated at that time that he suffered from lascivious dreams and emissions."[64] Since this patient's sexual desires were toward women, if she remained a woman, relationships she pursued would be considered same-sex and would perhaps be unacceptable to herself and to the doctor. Maybe the possibility of same-sex connections explains why Little was so frustrated with the patient for wanting to continue living as a woman and considering castration. That no such surgery took place suggests that Little was reluctant to castrate a patient he considered male, perhaps because he wanted to leave open the possibility for heterosexual intimacy. Little may also have been offended at the patient's resistance to his professional advice and at the patient's lack of enthusiasm for the male role. She did not want to be a man and rejected the value of masculinity by requesting voluntary castration.

In 1903, doctors encountered a patient who sought genital surgery in a controversial case that supported some doctors' preference for heterosexuality and challenged others'. E. C., a twenty-year-old "pseudohermaphrodite," arrived at Dr. J. Riddle Goffe's office in New York because a genital "growth," as she called her enlarged clitoris, was "a great annoyance." "It made her different from other girls, and she wanted it taken out," the doctor recorded. The growth was prominent: three inches in length and three and a half inches in circumference. Goffe complied with her request. Like Dr. Samuel Gross in the mid-nineteenth century, Dr. Goffe received harsh criticism, not because he performed genital surgery, but because some thought he had made the wrong call. Goffe had been convinced that the patient was a woman, despite heavy beard growth, thick eyebrows that met over the eyes, no breast development or menstruation, and a clitoris three inches long.[65] His colleague, Dr. Fred Taussig, suggested that what Goffe removed was not a clitoris at all—but that he had misread the patient's anatomy and symptoms and had excised "his" penis. Even the editor of the journal that published Goffe's account disagreed with Goffe's assessment, subtitling the article, "Operation for Removal of the Penis," when throughout the essay the organ was consistently called the clitoris.

Why didn't Goffe assume E. C. was male? The secondary sex characteristics seemed to point in that direction. Perhaps Goffe wanted to place the patient "safely in the ranks of womankind" because he had inquired into E. C.'s love life. He wrote, "She has never had any girl love affairs or been attracted passionately by any girl, but has been attracted by boys."[66] When Goffe responded to criticism of his surgical decisions, he elaborated on the relationship between the determination of sex in ambiguous cases and homosexuality. A crucial step would be to examine the ovarian or testicular structure, he admitted. But, as examining them would require removing them (thus unsexing the patient), the best alternative would be to "make a study of the individual mental and emotional attributes from a physio-psychological point of view."[67] It was well-known, he argued, that hermaphrodites born with a "duality of development" in their sexual organs turn out be "sexual perverts, or . . . inverts." And so, "the sooner they can be relieved of the duality and the anatomical features made to harmonize with the psychic the better it is for that individual and for society." In other words, it was necessary to alter patients' bodies so that their desires would be heterosexual.[68] E. C. needed to be a woman, in Goffe's eyes, because she had been romantically inclined toward boys. If Goffe had considered her clitoris to be a penis, then by classifying E. C. as male, the doctor would perhaps have encouraged same-sex relationships.

Goffe's case also provoked physician debate over how much respect to give a patient's own wishes. When Dr. Goffe had asked E. C. if "she preferred to be made like a man or woman, she said decidedly, 'a woman.'" Dr. Goffe performed the operation on March 11, 1903. In this case, the patient's desire—in both senses of the term, her own gender perception and her sexual attraction to men—matched the physician's propensity to make bodies align with heterosexuality. With "that thing" (as she called it) removed and her vaginal opening enlarged, E. C. could be a woman. In a follow-up visit the following October, she described successful electric depilation of her facial hair. Examination of her vagina revealed vaginal walls that were "smooth and satisfactory in every way." Goffe described her external genitalia as having "a perfectly normal appearance." As important, Goffe said that she was "in a buoyant frame of mind," and we are left to assume that she lived happily ever after. Goffe had a further opportunity to silence his critics when he saw E. C. seven months later; she reported that she had begun to menstruate. Thus, he wrote, there could "be no further question as to the propriety of the operation I performed."[69]

Goffe is unusual in the annals of intersex management, not because he wanted to guarantee heterosexuality, which was typical for doctors of the era, but because he explicitly asked the patient what she wanted and then complied with her wishes. As a result, he had to defend himself from critics who condemned what they saw as giving undue power to the patient. Taussig essentially accused Goffe of giving his patient the choice of which sex she would adopt; the title of his disapproving article in response to

Goffe's surgery was "Shall a Pseudo-Hermaphrodite Be Allowed to Decide to Which Sex He or She Shall Belong?"

Whose decision is gender? Goffe had his critics, but his conviction that people should make this decision for themselves echoed that of James Parsons back in 1741. As we saw in chapter 1, Parsons's guidelines read: choose male or female according to the "Predominancy of Sex, which ought to be regarded; but if the Sexes seem equal, the Choice is left to the Hermaphrodite."[70] But that recommendation had eroded during the nineteenth century as many doctors used their authority to rectify ambiguity and, especially during the late nineteenth century, to promote heterosexuality.

Into the twentieth century, doctors' reports betray the tendency to distrust their patients' words, particularly with regard to sexual desire. The tropes of the hermaphroditic monster and the deceiver lingered. In 1917, two leading New York physicians described a fifteen-year-old African American girl in terms that doubted her veracity while they highlighted the conflation of race, class status, and sexual uncertainty. Betty entered the hospital for "ulcerative affection of the external genitals," which she attributed to a rape four months prior. After five weeks of treatment with mercury and iodide, with the lesions sufficiently healed, doctors turned their attention to her atypical genital presentation and spent the next six months examining her physical and psychological health. "Betty is a dark negress of low intellectual type," they reported. "Her mental faculties cannot be called subnormal for the class and type that she represents." Betty said that she was attracted to men, but according to the nurse in the hospital ward, she paid no attention to the men, and instead "she is very devoted to the females in the ward, fondling them whenever permitted and unchecked."

The doctors were more inclined to believe the nurse's observations than Betty's words. In fact, they cautioned, "Her own statements have to be taken with reserve." Betty's accounts of her sexual experiences were inconsistent, the doctors believed; she admitted having been raped, but she also spoke of "her own sensations during intercourse." Adding their own assumptions about black women's sexuality, the doctors concluded that "her stories of rape by a white man only some months ago as her first sexual experience do not seem likely to be true in an individual of her race and age; sexual life usually begins much earlier." Perhaps Betty's presumed penchant for deceit should come as no surprise since, according to these doctors, hermaphroditism was thought to occur alongside other "mental" problems, including "hysteria, epilepsy, psychoses, criminal tendencies, and abnormal sexual inclinations."[71]

Betty's external genitalia appeared male, though one doctor who examined her considered "the penile mass as an extra clitoris and advise[d] its amputation." The primary doctors disagreed. They believed that Betty possessed both a penis and a clitoris, and though they could not call her a true hermaphrodite, "which is so rare as to be almost unrecorded in the literature of the subject," they suggested that she could hardly be any closer. They ultimately declared Betty to be "preponderatingly female because of the presence of a vagina and cervix, and in spite of the presence of a penis and of sexual impulses toward the female sex." As we have seen again and again, in their compulsion to choose one sex or the other, doctors confronted the very inconsistencies (a woman with a penis) they had hoped to avoid by insisting on a rigid binary of male or female. Doctors could see Betty not as a hermaphrodite, but rather as a "pseudo-hermaphrodite" though one with "bisexual external organs" and single-sex reproductive glands (in this case, presumed ovaries because she menstruated, albeit irregularly) and "psychic hermaphroditism," a reference to what the doctors believed to be her attraction to both men and women.[72]

In the early twentieth century, it was still all but impossible for a person to be labeled a true hermaphrodite. Many of the themes surrounding hermaphrodites endured from the colonial period, including the tendency to employ the term in denying the existence of what it denominated. The motifs of monstrosity, mistaken sex, deliberate deceit or fraud, and the potential for celibacy or nonheterosexual intimacy continued to capture doctors' imaginations and arouse their anxiety. But other notions, such as blaming mothers' harmful thoughts or questioning whether or not those born with unusual anatomies had souls, gradually receded.[73] The next chapter will examine the debate between patients' and doctors' wishes regarding treatment in the early twentieth century, as surgical and hormonal intervention became increasingly possible and popular.

CHAPTER FOUR

Cutting the Gordian Knot

Gonads, Marriage, and Surgery in the 1920s and 1930s

It is usual for the general public to think that a doctor can settle this question [the determination of sex] without difficulty. Experience, however, has shown that there can be few more tangled Gordian knots presented for unravelment than this same question of the determination of sex in certain doubtful cases . . . Even when such cases come to autopsy it is extremely difficult to decide whether the sexual organs found are ovaries or testicles, and, of course, on these organs depends the essential distinction of sex.

A S THE EPIGRAPH FROM 1898 suggests, there were no easy ways for patients and their families to determine the "essential distinction of sex," though most doctors remained convinced that the presence of either ovaries or testes would provide the answer. By the first half of the twentieth century, doctors were able to assess the gonads by taking tissue samples and examining them under the microscope, and so a person's sex could presumably be ascertained while he or she was still alive. Sometimes, though, a person's visible genitalia contradicted the biopsy. Then doctors faced important decisions. They held information both recondite and, by their criteria, determinative, and so they faced ethical and moral dilemmas. Should they surgically alter external genitals to match internal gonads? Should they encourage patients to change their lifelong sexual personas? Should they even disclose what they found? This chapter explores how doctors answered those questions as they confronted patients whose mixed nature challenged both medical ethics and surgical skills. In the 1920s and 1930s, as surgical techniques grew more sophisticated, surgeons increasingly thought they could "make" men and women. Weighing the trials and failures of earlier surgery against their own presumptions of competence and their own concepts of socially acceptable bodily integrity, they ultimately justified genital surgery by social, rather than purely medical, goals.

The Gonadal Standard

In the 1920s a woman came to Dr. Leon L. Solomon of Louisville, Kentucky, seeking an abortion. He refused to perform an abortion but agreed to do a physical examination. What he encountered in the examination room stunned him: "Upon entering the room, to my amazement there stood, nude, a veritable wolf in sheep's clothing, with all of the outward habiliments of a man. I felt that a hoax was being played on me and am free to admit, I did not know whether I was angry, frightened or embarrassed. Gazing first into a woman's face, then at a large male organ of procreation was sufficient to produce a queer sensation."[1] Was this person male or female? Her pregnancy implied womanhood, but the large phallus suggested otherwise.

Dr. Solomon admitted uncertainty as to whether his patient should rightly be called a woman or a man. Perhaps because of this ambiguity, he resorted to the kind of descriptors long used to label people with atypical genitalia; he called her a "strange creature," an "unfortunate creature," a "queer individual," and "a strange fellow," highlighting the mixed nature of her body. Like most women, she had no beard on her face, but she had heavier hair on her arms and legs than other females. Her voice was "usually soft and mellow," but at times it was "more masculine than feminine." Her hands were large, implying masculinity, but her hips were "distinctly feminine"; her lower legs were "not distinctly feminine, but her feet were small." Consistent with her pregnant status, her breasts expressed milk, a further indication of womanhood, even though she also had a penis. The doctor thought she had testes as well, but he was unable to positively establish this because she would not let him palpate them. "Manipulation of these two bodies is resented," he reported, "on account of sensitiveness." He found a vagina and assumed she had ovaries because of the pregnancy. Dr. Solomon published an account of his examination of this patient entitled, "Hermaphroditism: Report of a Case of Apparently True Hermaphroditism with Photographs of the 'Woman.'" The title suggested that Dr. Solomon believed his patient to be a "true hermaphrodite," which he defined as a person "who possesses male and female sex organs, described in the literature as Double-sexed or Complex Hermaphrodism."[2]

After she left Solomon's office, the woman ended the unwanted pregnancy on her own, using a catheter to abort the fetus, as she had done successfully in instances past. Solomon saw her again a year later, and he noted her visible deterioration. She had become promiscuous; she dressed in women's clothes and had heterosexual sex as a woman but regularly masturbated as a man. Before she had struck Solomon as a woman "possessed of some refinement"; now she "gave evidence of a vulgar, immoral woman." Perhaps Solomon had grown disenchanted with the patient because of the abortion or because of revelations of her sexual activity. Solomon interpreted her behavior sympathetically however, proclaiming that she was "more sinned against than sinning."[3] Altogether perplexed, Dr. Solomon seemed to throw up his hands in despair, proclaiming, "Verily, an All-wise Creator sometimes deals with His handiwork in a manner, beyond mortal understanding."[4]

Solomon's befuddlement was somewhat unusual for the time period. For explanations he could have turned to a then-popular theory about sex development that relied on cell development, rather than on God's incomprehensibility: the theory of "bisexuality." The concept had to do not with sexual attraction, as the word does today, but rather meant that at some stage of its growth, each developing embryo had the potential to develop into either a boy or a girl. Each embryo was hermaphroditic, the theory alleged. According to William Blair Bell, founder of the College of Obstetricians and Gynaecologists in London, "There must be a latent but predominating tendency in every fertilized ovum towards masculinity or femininity, and . . . every fertilized ovum is potentially bisexual."[5] In the 1930s the theory still held sway. Emil Novak, an eminent Chicago gynecologist, explained, "In short, there are few 100 percenters among human beings from this standpoint, there being a bit of the feminine in all men and a corresponding tinge of the masculine in all women."[6] But at some point in embryonic development, one set of sex organs would begin to develop, while the other atrophied. Hermaphrodites, then, were those in whom the process went awry.[7]

The theory held that sexual development progressed along a defined path and that ultimately each individual developed a pair of gonads that determined the person's sex. "What do we really mean when we speak of male and female sex?" asked Dr. D. Berry Hart in 1915. "The only criterion of sex is the nature of the sex-gland. In the male we have a testis containing sperm-cells . . . ; in the female, an ovary containing the non-motile oocytes or ordinary ova." According to Hart and others, "When we have a certain disturbance in the structure of the sex gland (i.e., the coexistence of sperm-cells and oocytes) we have true hermaphrodism." [8]

As Alice Dreger has noted, while medical experts in France and Britain touted the reliability of the gonads to define true hermaphrodites, and thus looked for both ovarian and testicular tissue in a single person, the "application of the gonadal dictum in practice was at best uneven."[9] The historian Alison Redick agreed that, in this country too, the gonadal standard fell short during the early twentieth century; she has argued that the period is more accurately labeled the "era of idiosyncracy" rather than the "age of gonads." Redick is right to note that doctors did not rely solely on gonads to define sex, but their idiosyncratic approach to patients exhibiting mixed genitals and secondary sex characteristics is not unique to this period.[10] In fact, a diverse and seemingly random approach typified doctors' reactions to such patients since the seventeenth century, as we have seen. As the twentieth century progressed, new technological and scientific advancements made it possible to evaluate new factors: gonadal tissue, hormonal levels, and even chromosomes.[11] But none of those indicators could (or can, even to this day) tell doctors with absolute certainty what the "actual" sex of an individual is or whether a particular person would identify as the sex the doctors decided on. With the definition of a true hermaphrodite undergoing revision with each technological advance, what remained consistent throughout the first half of the twentieth century was doctors' commitment to heterosexual marriage and, increasingly, to surgery devoted to guaranteeing the union of two differently sexed bodies by the creation of "perfected" men and women.

Social Justifications for Surgery

The justification for risky surgeries consistently emphasized social over medical concerns. In a 1926 article in the British medical journal the *Lancet* regarding hypospadias, Dr. Arthur Edmunds spoke to a transatlantic audience in his rationalization of surgical activism: "I venture to think that there is no deformity for which treatment is more urgently called. For example, a lady of my acquaintance had a little son born with this deformity. Her brother, who was a medical man, regarded it as a family calamity, and a medical friend expressed the opinion 'that it would be a good thing if the child died.'" Believing that the child would be victimized on the playground, Edmunds explained that "the parents were extremely anxious that his deformity should not be known. He was not allowed to go to school or to go to any parties or festivities, where the need to urinate and the posture he was compelled to adopt would reveal his condition." According to Edmunds, the child underwent repair and emerged unscathed from the ordeal. "The child now mixes ordinarily with other children and is preparing for a public school," he proudly declared. [12]

Edmunds recognized that his primary motivations for surgery were social rather than strictly medical. He mentioned the tendency to chronic wetness in the genital area if the urine stream could not be projected forward, but sitting rather than standing to urinate could presumably alleviate that problem. Edmunds emphasized what he believed to be the more significant justification: "The child himself realises that he is different from the rest of the community and is liable to various forms of perversion. Marriage is, of course, impossible, and in every way the child's happiness is seriously handicapped." [13] For many doctors, worries about childhood social isolation merged with anxieties about sexual perversion and the inability to achieve heterosexual penetration. The potential for successful marriage (or, more specifically, heterosexual intercourse) hence came to justify genital surgery, regardless of the patient's age or interest in marriage. [14]

As early as 1905, Charles-Marie Debierre, author of a prominent medical textbook translated from French into English, argued that it was justifiable for a surgeon to create a vaginal cavity when a woman wanted to accommodate penetration during intercourse. He argued that if a woman wished surgery, "either for the satisfying of the sexual passion, and to retain her husband if she is married, or if she intends to marry, then it is the surgeon's duty to acquiesce." All women, he reasoned, even those with "sexual infirmities," would want to marry: "What would, indeed, become of the unfortunate woman, who, perhaps, is excited by her husband with a passion which he is unable to gratify? For it is not reasonable to suppose that women who are afflicted with this malformation have not the desires and hopes of their sex."[15]

Indeed, doctors insisted that it was their patients' desires for "normal" heterosexual sex that motivated vaginal lengthening and clitoral excision.[16] Dr. J. Mark O'Farrell of Houston, Texas, saw three sisters with similar genital anomalies in 1935. All three were "clean minded and there is no suspicion of perversion." In other words, he suggested all were sexually attracted to men (though one was only nine years old) and therefore wanted their enlarged clitorises removed and their vaginal cavities lengthened in anticipation of marriage. The middle daughter, nineteen years old, "demanded," in fact, that "the penis be amputated, failing which she intended to commit suicide." Finding no testes and concluding that she was indeed female, O'Farrell tried "to improve the vaginal outlet." Considering future sensation, he amputated what he called her penis, "with enough left at the base to provide a possible substitute for the clitoris, and with the idea in mind to protect the urethral canal."[17]

He described the success of the surgery on the nineteen-year-old:

The attitude of the patient was immediately improved. She became optimistic almost overnight. At every visit to the office she had some new feature of improvement to report, and, at times her enthusiasm led her to exaggeration. However, her development has been little less than sensational. She no longer uses depilatories. The beard has practically disappeared and the face is much smoother. The breasts have developed to the size of small lemons, and the nipples are prominent. Her voice is less masculine but not yet as satisfactory as desired. She is quite satisfied with her condition; she attends social functions and appears interested in the young men of the neighborhood.[18]

In essence, she became more womanly: less hair, more breast growth, a more feminine voice, and greater sexual interest in men.

The older sister was not so heterosexually inclined as the middle daughter, but O'Farrell concluded that she, too, was a woman because she had normal ovaries and no testes were found. This woman, twenty-four years old, "was tortured with extreme sexual desire, without the ability to determine the preferred sex." She considered masturbating but restrained herself. She thought of herself as a mule, "without hope or promise beyond a day's labor." Though raised female, she believed "that the male characteristics predominated, especially since the penis appeared to be making increasingly insistent demands for sexual satisfaction." Perhaps O'Farrell believed that the same womanly qualities that transformed her younger sister would appear following the surgery, though he did not include an explicit assessment.[19]

The youngest girl, aged nine, had a feminine face and voice and no testes. She had what O'Farrell considered a penis, "much larger than is ordinarily found in a boy prior to the age of puberty." After it was removed, this daughter was also a success story in O'Farrell's opinion. He wrote that now she was "apparently a normal girl." Consistent with the trend toward secrecy that will be discussed below, he recommended that "her past sexual anomalies never be discussed with her."[20]

Marriage

In the early twentieth century physicians strove to make matrimony part of their professional domain. In earlier eras doctors had not typically been consulted when two people considered tying the knot. As the historian Wendy Kline has argued, doctors' assertion of expertise on marriage and childbearing decisions reached its peak with the popularity of the positive eugenics campaign of the 1940s and 1950s.[21] But as early as the 1910s, marriage manuals, often written by physicians, included chapters on doctors' influence over marriage suitability. William Robinson, author of several such guides as well as a medical treatise on sexual impotence, explained that "it is now a very common occurrence for the intelligent layman and laywoman, imbued with a sense of responsibility for the welfare of their presumptive future offspring and actuated, perhaps, also by some fear of infection, to consult a physician as to the advisability of the marriage, leaving it to him to make the decision and then abiding by that decision." Couples with various hereditary diseases were advised to avoid marriage and pregnancy, and doctors warned that even conditions where hereditary influence was inconclusive, such as hysteria or alcoholism, should be investigated thoroughly before parenthood could confidently be recommended. [22]

Doctors were of two minds regarding hermaphrodites and marriage. They generally wanted to ensure that their patients' bodies were properly equipped for heterosexual penetration, should they indeed marry. But some were convinced that hermaphrodites should not marry at all, given their ambiguous condition. One anonymous writer in the *Medical Record* had explained back in 1892, "Socially an unfortunate object, and much to be pitied, the hermaphrodite belongs to the dangerous classes, and society should be warned accordingly. Humanity is either masculine or feminine . . . Before the civil code it would be well to consider the hermaphrodite absolutely neuter, thus excluding the possibility of legal marriages in cases of doubtful sex."[23]

Though they may have preferred that hermaphrodites not wed, doctors were surprised to find that many did, sometimes as the "wrong" sex. A person might have been raised as a girl, lived as female, married a man, and yet find out years later that her body contained testicles rather than ovaries. Thus, in the minds of many physicians who accepted the gonadal definition of sex (it is what you have on the inside that counts), a person who fit that description was really male, not female. Dr. Charles Creevy of Minneapolis, for example, lamented in 1933 that "most external and complete pseudohermaphrodites are reared and marry (if at all) as members of the wrong sex."[24] Dr. Milton Helpern of New York described the case of a thirty-seven-year-old African American man who died of alcoholism. He had been raised as a boy, lived as a man, and had been married for twenty-two years. Yet this man, according to the doctor's report, had feminine breasts, soft skin, no beard, and a hypospadic penis. After he died, doctors performed an autopsy and found a uterus, cervix, vagina, and ovary on one side, and an *ovotestis* (a gonad that contained both testicular and ovarian tissue) on the other. Because of the ovotestis, this person was classified as a "true" hermaphrodite. Often considered impossible and nonexistent, such an individual was generally recognized as the rarest of all hermaphrodites. When the doctor questioned the dead man's wife about their life together, she reticently told him that their sex life was "fine" but that she had never been pregnant, thus confirming the doctor's suspicions that, although he had lived his life as a man, this person was not actually a man because he had never impregnated his wife.[25]

Since he believed that this patient was not male, indeed, more a woman than a man, Helpern asked the wife about any "cyclic behavior" she might have noticed while her husband was still alive. It is obvious from Helpern's tone that he had misgivings about the wife's reply: "As far as I could learn from his wife," he wrote, "who was rather reluctant to talk about him and who seemed to be very naive about the matter, the man seemed masculine in his traits, and as far as his libido and his conduct during coitus were concerned, she said he seemed to be all right! During his alcoholic 'sprees' she knew very little about what he did or where he went." Perhaps disappointed with the wife's nominal responses, Helpern concluded that since the man's internal organs were those of a woman, he must have menstruated, though to Helpern's regret, he "could not obtain any information about this phenomenon from his wife."[26]

The Ethics of Surgery

The possibility of people living in the wrong sex worried physicians and dictated how they interpreted their patients' lives and decided on their treatment. Of course, if "true" sex was discovered at autopsy, no further action could be taken, but, for living patients with confused genitals, most doctors in the 1920s and 1930s promoted surgical and hormonal intervention. One doctor explained in 1923, "It has happened that certain unfortunate individuals have passed part of their life under the belief that they belonged to one sex only to discover that strictly speaking their sexual organs, as they develop in full at maturity, proclaim them to belong to the opposite sex. It is as if Nature did not complete her work of differentiation in the development of the sexual organs, as if she hesitated midway and left the work only half done."[27] Doctors would finish what Nature had suspended. In presenting their cases they stressed the deceptions and misjudgments (intentional or not) that were uncovered as well as the happiness felt by all concerned when things were righted. The move toward surgical "correction"—often an inaccurate term since surgery does not cure many intersex conditions—intensified the tendency to glorify and dramatize outcomes.

Hugh Hampton Young, the premier urologist at the Johns Hopkins University in the first half of the twentieth century and author of the leading 1937 textbook *Genital Abnormalities, Hermaphroditism and Related Adrenal Diseases,* recognized that because of intersex conditions, people frequently lived as what physicians considered the "wrong" sex: "In such persons the sex often has been wrongly interpreted at birth, and they have been brought up as boys when really girls, and vice versa. As the child grows older, the parents come to the physician filled with anxiety as to its sex. In many cases abdominal section and examination of the pelvic organs must be carried out before it is possible to determine the true sex of the patient."[28]

Doctors tried to discern the "true sex," so that the correct course of action could be pursued, but this search proved elusive even as medical knowledge refined. By 1937, when Hugh Hampton Young published his tome, a volume that included 55 case studies, 379 photographs, and 534 drawings, the nomenclature had been established, but the treatment options were still much contested. The labeling was confusing for patients, parents, and even doctors because the label focused on the gonads, thus often contradicting outward physical appearance. If the person had female external genitals but internal testes, she may have presented herself as female, but doctors labeled her a male pseudohermaphrodite. If a male patient had a phallus but internal ovaries, then he was called a female pseudohermaphrodite.[29] To make matters worse, labeling did not suggest treatment for hermaphroditic patients. Though Young's book presented the most up-to-date and sophisticated surgical techniques, it nonetheless offers readers a disheartening journey into the lives of men, women, and children who came to doctors for help and were often left in worse shape, physically and mentally, than when they arrived.

In the 1920s and 1930s, frankly, doctors experimented on *all* patients' bodies, as they theorized about what surgery to perform and how to counsel and guide their patients.[30] Surgeons themselves came to look back on the period with dismay. C. Everett Koop began his tenure as surgeon-in-chief at the Children's Hospital of Philadelphia in the 1940s, and he later lamented, "none of the correctable congenital defects incompatible with life had ever been successfully treated in Philadelphia except rarely and then by good luck more than good management. The mortality for a simple colostomy was in the range of 90%."[31] Performing surgery on children also posed unique challenges. Doctors had not yet established pediatric surgery as a separate discipline; general surgeons or anatomical specialists, in this case urologists, performed most procedures. It was not until 1941, with the publication of the first modern textbook on pediatric surgery, that doctors began treating their young patients as children, with their own set of medical and pharmacological needs, rather than as miniature adults.

Of course, doctors believed that they were helping their genitourinary patients, even when they were unconvinced that their procedures would be successful. In their minds, they were fashioning complete men or women out of "unfortunate creatures," a commonly used term that epitomized their paternalism and distaste. And so with the lofty goal of creating "real" men and women, they tried one surgical procedure after another, without thoroughly considering the medical or the psychological risks of the operations. Should a person raised male, but whose body contained ovaries, be allowed to continue living as a male? Should doctors remove the penis of such a person because they believed it to be not a penis but a large and embarrassing clitoris? Who was authorized to make the decisions, and how would one know the right decisions had been made? What constituted a positive outcome?

We cannot reasonably hold the past to the standards of medical care that we expect today, but we can be critical of the overly optimistic accounts that doctors published of their surgical procedures. Today, patients and doctors are demanding that medical care be based on a systematic analysis of what has worked and what has failed in the past (evidence-based medicine), but in the first half of the twentieth century, the data were simply not available to offer patients the information necessary to give adequately informed consent.[32] Even when doctors admitted the limitations of one surgery or another, they often pushed ahead with it.[33] In a 1922 discussion about the treatment of hypospadias, for example, the physicians Francis R. Hagner and H. B. Kneale acknowledged that "undoubtedly more unsuccessful operative attempts have been made, that have not been reported, than successful operations that have been reported." Yet despite this honest disclosure, they believed that such surgery would have a social benefit. When evaluating what to do with a specific patient whose hypospadias precluded heterosexual penetration and required the patient to sit while urinating, the same doctors wrote, "Such a sufferer can, indeed, bless the hand that reconstitutes him a man."[34]

Doctors' surgical decisions were biased by a wish to do *something*, perhaps by an underlying feeling that, should surgery fail, a genitally indeterminate person was better off dead, and by their own subjective evaluations. Commenting on fourteen of Hugh Hampton Young's cases presented at a meeting of the New England branch of the American Urological Association in 1933, Dr. William Quinby admitted, "It is exceedingly hard to decide what action one should take as a physician when confronted with an individual of this sort, and it is probably a matter which will be decided, in the last analysis, according to the characteristics of the individual."[35] But how doctors interpreted those characteristics varied according to whether they saw that individual as male or female. Sometimes doctors ignored the personality and the wishes of the patient, as they trusted their own judgment. The following case illustrates how, once committed to a decision about a patient's "true sex," doctors *saw* that individual as male or female, despite both evidence and the desires of the patient to the contrary.

A seven-year-old boy was admitted to the Johns Hopkins Hospital in October 1925, diagnosed with undescended testicles and hypospadias. The child had been raised in an orphanage as a boy, even though authorities knew that he was physically "abnormal." In addition to a penis, the boy had a short vagina and no visible testicles, though doctors thought that they might be undescended and still in the abdomen. Had they found testicles, they would have repaired the hypospadias and left the child to live his life as male. What they discovered instead, though, was a "normal-looking ovary" on one side. As the presence of internal reproductive organs took precedence in deciding the true sex of a person, the report read, "it was decided that the patient was definitely female, with vagina, uterus, left ovary, two Fallopian tubes, no right ovary, testicle or ovo-testis being found, and with marked enlargement of the clitoris, simulating a penis." What was once seen as a boy's penis when a physician first examined the child was now seen as an enlarged clitoris. The patient was now read as a girl because of the internal organs.

The child's custodians were notified that the child was, in fact, female and accordingly, his name was changed from Frank to Frances. Though a gender reassignment had been effected, no surgery had been done. Two years later, Frances returned to the hospital. Now living as a girl, she complained that the "clitoral appendage" was annoying to her and that she felt pressure to keep it hidden. This made sense to the doctors, because most girls do not have large clitorises that need to be covered for fear of exposure. Doctors also mentioned that Frances had a "terrible habit" of masturbation. So in March 1927 doctors removed the clitoris, which they mentioned looked like a large penis, the size of that of a ten-year-old boy. Consonant with their decision that Frances was definitely female, the clitoris/penis was removed; no longer would Frances, the girl, be able to masturbate or perhaps experience any sexual sensation at all.

Four years later, when Frances was sixteen, she returned to the doctor, complaining of a mass in her lower abdomen. She was still living as a girl at the orphanage, but she announced that she planned to "assume male attire" when she turned eighteen and learn a mechanical trade. Upon learning of Frances's plans to live as a man, the doctors began to see Frances differently, admitting certain elements of masculinity that they had hitherto ignored when convincing themselves that Frances was female.

Doctors mentioned that "the patient was 5 feet, 11 inches tall and weighed 125 pounds, and typically male in stature . . . There was a rather thick growth of short, stiff, black hair over the chin. The mammary regions were entirely masculine. The pubic hair was female in distribution." Now that Frances wanted to live as a man, doctors suspected that the amputated organ might have been his penis. Further exploratory surgery revealed testicular tissue, indicating that, indeed, by their standards, Frances really was a young man. The damage had already been done four years earlier, with the removal of the penis; that surgery had made Frances female, according to the medical team, and so now they excised all the testicular tissue that they found and a rudimentary vas deferens. Even with all traces of maleness surgically removed, by 1935 Frances had changed his name to John and was working as a truck driver. The medical report noted, almost incredulously, that "he is apparently happy although living as a male without penis or testicles and with the vagina still present."[36]

As the previous chapter indicated, there was profound disbelief that men could live successfully without penises or women without penetrable

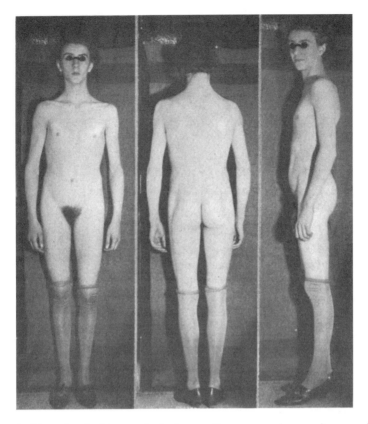

Though this patient had been raised a boy until age seven, doctors discovered an ovary. This discovery persuaded them that he should be a girl, and so they surgically removed what they thought was an inappropriately large clitoris. A year after this picture was taken, Frances changed his name to John and began living as a man, despite the earlier maiming surgery that had amputated what would now have been considered his penis. Hugh Hampton Young, *Genital Abnormalities, Hermaphroditism and Related Adrenal Diseases* (Baltimore: Williams and Wilkins, 1937), 89. Courtesy New York Academy of Medicine.

vaginas. Here was a person who seemed to defy those doubts. The last sentence of the case report betrays Young's skepticism. Perhaps still believing that this patient would be better off female, Dr. Young speculated, "If implantation of ovaries ever becomes successful in the human should this be done?"[37] Even though Dr. Young admitted that John's "mental outlook had been completely transformed," and that he seemed happily male,

Young hypothesized about ovary implantation, which presumably would have necessitated transforming John back into Frances against his wishes. The mistake of the excised penis would then be negated, and the doctors' initial decision that John was a woman would be supported.

Even though John no longer had a penis, his internal evidence of maleness guaranteed him the privilege of declaring himself male, despite the doctor's doubts as to his happiness.[38] Those patients whose gonads contradicted their genitalia as well as their sense of themselves faced a more difficult dilemma when confronted with new medical information. What made one a man or a woman? Did the gonads tell the whole story? Doctors too faced questions. Should they tell patients what they had found inside the body? And what, if anything, should be done with knowledge that challenged how people had been living their lives?

These questions were not merely rhetorical. With advances in histology, the science that studies the microscopic structure of organic tissues, doctors were able to take samples of gonads and ascertain exactly what kind of cell tissue constituted the organs. Even if the glands had not reached their full stage of development, photomicrographs of ovarian and testicular tissue could reveal different kinds of cell structure, which offered basic clues about glandular growth.[39] But evaluating patients based on their glandular tissue proved problematic. Even the previous case could not stand up to the gonadal standard because, though the young man's testicles were removed in order to correspond with his amputated penis, he still considered himself male. The contradiction between an individual's sense of gender identity and a strict adherence to the gonadal definition of sex sometimes dramatized the limitations of medical definitions of "true sex" and the way attachment to prevailing definitions could undermine ethical responsibility to patients.

In 1915, for example, Dr. William Quinby saw a ten-year-old boy who "spurned girlish pursuits" in favor of football and who, like his father, had hypospadias.[40] When Quinby first performed an exploratory laparotomy (abdominal surgery), he could not find any spermatic cord and instead found a uterus and ovaries. He told the parents that the child they had raised a boy was actually a girl. Quinby stood by the gonadal definition of sex. "The sex of an individual," he argued, "must always be determined by the nature of the gonad, regardless of the presence of abnormalities either of other parts of the genital system or of the secondary sexual manifestations of the body as a whole. Consequently, this patient is of the female sex; and this in spite of so many secondary characteristics of the opposite, male, sex."[41]

The boy's parents did not subscribe to the gonadal definition of sex, and they insisted that Quinby keep the boy's internal gonadal status a secret. Instead, they wanted Quinby to make their child even more of a boy via surgery by fashioning a urethra through what they considered to be his penis, presumably so that he could stand to urinate. Five months later, Quinby carried out the parents' wishes. The operation was unsuccessful. Three years later the boy needed another surgery to repair a fistula, or tear, in the new urethra, a common problem with urological surgery. The following year, when the boy was fourteen, another fistula was found when he returned for hypospadias repair. After two and a half months in the hospital, where he contacted severe dermatitis from the x-ray treatment required to remove the hair from his pubic region, the child returned home, only to return five months later with his sores still unhealed.[42]

It was unusual that this boy and his parents refused sex reassignment surgery, the creation of female external genitalia to match his internal reproductive organs. But, sadly, they were unable to escape the medical establishment's notion that the gonads dictated one's true sex. Seventeen years later, the patient returned to the hospital, to straighten out this business of his sex. He was thirty-one years old by this time and, according to Quinby's colleague Dr. Young, a "well formed man with a heavy beard." He owned a mercantile business and wanted to get married. Young described the man's fiancée as "a normal, good-looking young woman [who] professed to be much in love with the patient." Though his hypospadias had never been fully repaired, the man "admitted that he and she had been intimate on numerous occasions, but that his conformation did not deter them in their sincere wish to marry." He said that even though his penis was short, he had "excellent" erections and that intercourse was "entirely satisfactory." [43] What was the problem? Apparently, twenty years earlier, his father had mentioned to his priest that the child had ovaries rather than testicles. The priest considered him female, all outward appearances notwithstanding, and would not consent to or perform the marriage. The man pleaded with the doctors to declare that twenty years ago a mistake had been made and that he was indeed male.

The young man's heartrending appeal centered on desire, something that doctors took into consideration when evaluating adult patients but that had not come into play when this boy had been just a child. He implored, "I feel certain that with my strong desire for intercourse with females, and the fact that the act is apparently normal and entirely satisfactory to me and them, that a mistake must have been made in declaring me to be female, and that I am really male. My heart-felt desire is that this be proved so that my priest will consent to marry us."[44] Unfortunately, another round of examinations confirmed the initial findings from years before, and the doctors refused to overturn their previous decision about this man's sex, insisting that ovaries instead of testicles made him female, despite everything else about his appearance and his sense of himself as male. Forbidden to marry his intended bride, the man killed himself by ingesting bichloride poison.

So many questions emerge: why didn't the man seek another priest, one who didn't know his medical history? Why did he go back to the doctors at all? Maybe the man could not believe the doctors' original diagnosis when he so clearly felt himself to be male. Why couldn't the doctors have admitted uncertainty, especially given the man's social presentation and obvious success in the ways heterosexuality was usually calibrated? After all, he functioned well socially, owned a mercantile business, and was engaged to be married. And for many doctors, heterosexual fulfillment not only justified previous surgery but signified success. But gonads were supposed to tell the truth. As the man's gonads were ovaries, his doctors and his priest considered his true sex to be female, despite all indications to the contrary, including the man's own sense of himself as male.

The Ethics of Uncertainty

Doctors debated the best course of action for patients whose gonads contradicted their genitals, recognizing the difficulty of changing one's gender presentation. If the patient was a child, then doctors presumed it would be easier to switch. No sexual attraction would complicate matters, and the patient's gender identity was presumably not yet cemented. In 1930,

Dr. John B. Carnett wrote about a five-year-old child who had been raised a girl but who was, the doctor thought, really a boy. At birth her mother had noticed an "abnormality" but thought the infant was female. Over time, the "small fleshy tab" grew and eventually, the "penis-like ornament has begun to excite comment by girl playmates." The doctor initially thought that the child had hypospadias and undescended testicles, and so he told the parents that "there was no reasonable doubt as to the child being a boy and that henceforth he should abandon the long hair and dresses of childhood."[45]

The parents were not convinced and sought reassurance with an exploratory laparotomy. They told the doctor that if he found both testes and ovaries, they wanted the testes removed so that their daughter could continue living her life as a girl. Surgeons found a uterus, tubes, and ovaries, though all were small even for a five-year-old, and no sign of testes. Carnett wrote reluctantly, "We are therefore compelled to revise our diagnosis and now beyond all doubt declare this child to be a girl. In other words, she is a pseudohermaphrodite of the gynandrous type." Finally, he wrote, "To prevent future embarrassment I will also amputate the over-developed clitoris just beneath the skin level and close the skin over it."[46] First Carnett had attested that "there was no reasonable doubt" that the child was a boy; later he flipped and said that the child was a girl "beyond all doubt." His upended certainty reveals a deep truth about intersex management in the 1920s and 1930s: physicians were not sure what to do with their patients, but only a few would admit insecurity.

Some doctors resisted the trend of professed conviction and confessed doubt about the efficacy of intersex treatment based on the gonadal system. Dr. James F. McCahey wrote, "Present day management of hermaphroditism is based on this explanation, but the therapy is unsatisfactory because of uncertainty as to the proper procedure in any given case."[47] McCahey argued that if one looked closely at so-called ovaries in hermaphrodites, one would find "male elements." The gonads did not tell the whole story, he reasoned: "Sex can not be identified with the nature of the sex glands, since these may be abnormal in females."[48]

McCahey's objections recalled a twenty-year-old dissent from the gonadal standard. William Blair Bell, a British author cited frequently in the United States, had argued in 1915 that it did not make sense to rely on the gonads to determine sex. "It used to be thought that a woman was

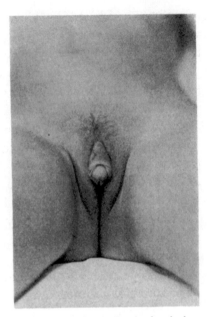

Fig. 543.—Hypertrophied clitoris simulating a penis.

The parents of this little girl did not believe their doctor's assessment that their child was a boy, despite the appearance of the genitals. After internal exploration found rudimentary female reproductive organs, the doctor reluctantly agreed that the child could remain female. J. B. Carnett, "A Case of Gynandrous Pseudo-Hermaphroditism," *Surgical Clinics of North America* 10 (1930): 1325. Courtesy New York Academy of Medicine.

a woman because of her ovaries alone," Bell reasoned. "But . . . there are many individuals with ovaries who are not women in the strict sense of the word, and many with testes who are really feminine in every other respect."[49] Bell pointed out that how people lived their lives did not always match up with their gonads; he contended that "the psychical and physical attributes of sex are not necessarily dependent on the gonads." Each case, he postulated, had to be considered individually, emphasizing the "obvious predominance of characteristics, especially the secondary, and not by the non-functional sex-glands alone, for this is neither scientific nor just." McCahey reprised Bell's argument that secondary sex characteristics and psychological characteristics ought to have some bearing on the determination of sex, gonads notwithstanding.

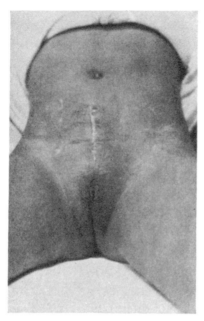

Fig. 544.—Postoperative appearance

With her enlarged clitoris removed, this girl appeared "normal" after surgery but no doubt sacrificed sexual sensation, a factor that her doctor did not take into consideration. J. B. Carnett, "A Case of Gynandrous Pseudo-Hermaphroditism," *Surgical Clinics of North America* 10 (1930): 1327. Courtesy New York Academy of Medicine.

Relying on the gonads to determine a patient's true sex was becoming increasingly debatable, but without an alternative solution, doctors in the 1930s still clung to the notion, if ambivalently. In 1933 Charles Creevy asserted, "All treatment must be based upon accurate knowledge of the actual sex. As has been pointed out, this depends entirely upon the histology of the gonads, and may be secured only by biopsies." In the same article, however, he retreated from this declaration. Though "the sex gland is, in the last analysis, the practical determinant of sex . . . , [i]n certain cases it may be advisable to disregard the sex glands if the secondary sex characteristics, the external genitalia and psychosexual outlook are all opposite to them."[50]

Few physicians admitted uncertainty, but some, like Dr. Emil Novak, a gynecologist at Johns Hopkins, acknowledged that they did not have all the answers on this subject. Perhaps its very ambiguity was what made Novak proclaim in 1935 that "there is no more interesting biologic or clinical problem than that of intersexuality." He eschewed confidence in the gonads. "Even the character of the gonads, on which the decision of actual sex has been commonly based in doubtful cases, is an incorrect criterion, as the biologic studies of recent years have shown." Then again, the external genitalia were not "safe criteria" either, he pointed out, because "typically female external organs have been found in individuals in whom the gonads, and perhaps the only gonads, were testes."[51] Relying on any one indicator was inconclusive.

Novak was just as interested in arriving at a treatment plan, despite all contradictions, as he was in pinpointing the exact determinant of sex. One of his cases concerned a young woman in college. Her first symptom was failure to menstruate. Without elaborating, Novak reported that she felt "different" from other girls, but that as a child she had played with dolls and had a maternal instinct. Upon examination, he found that she had no uterus or ovaries and that her gonads had "testicular elements." If Novak had adhered to the gonadal standard, he would have labeled this young woman a man. But it was clear to Novak that she did not identify as a man. Novak performed the "feminizing procedure" of removing the young woman's enlarged clitoris. It is ironic that doctors considered it feminizing to remove a woman's clitoris; after all, a clitoris is uniquely female. It was not the organ per se that was problematic, but rather its resemblance to a penis that persuaded doctors like Novak to amputate it.

Some doctors recommended removing problematic gonads, in addition to enlarged clitorises, in order to allow patients to resume their lives with fewer complications. In 1933, Dr. Louis Cohen of St. Louis wrote a paper in which he agreed with Bell's and Novak's position that the gonads alone were inadequate to determine sex. His patient, a twenty-year-old woman, had been raised a girl yet developed male secondary characteristics at puberty. Most notably, she "developed an intense sexual urge directed almost exclusively toward the male sex. The urge has been at times so intense that nothing has stopped her in an attempt at gratification." In fact, her parents had to institutionalize her "not only to keep her from inviting men, but actually from assaulting them." Upon internal inspection, doctors found functioning testicles, which they removed. Cohen explained, "This was done in order to remove the overwhelming and overpowering sex urge and with the hope that castration would cause an accumulation of fat about the pelvis and in the breasts, stop the growth of hair on the face, and allow the voice to become higher pitched." Since his patient had female external genitalia, "the addition of feminine secondary sex characteristics will allow this patient to continue in society as a female."[52]

Novak similarly insisted that even though the gonads of his patient, the college student, were those of a man, "there seemed to be no question whatever that the patient should be allowed to continue life as a female . . . Aside from her dominantly female characteristics and her female psychology, the external genitalia were of the normal feminine type except for the overgrown clitoris and her rudimentary vagina."[53] Her secondary sex characteristics and her sense of herself as a woman outweighed her hidden testes. Novak even saw her ambiguous genitalia as essentially manageable, nothing that a little surgical procedure could not fix.

Novak's case raised an important ethical dilemma for physicians: now that they were able to assess gonadal tissue that contradicted external gender presentation, should they encourage a change in sexual persona? Should they even tell patients what they had found, particularly if they thought the patients would be better off staying in the gender in which they were raised? In today's world, these questions are more easily answered or at least more openly addressed. Contemporary intersex activists have fought for doctors to treat intersex as they would any other health issue, with full disclosure to the patient or to the parents, if the patient is a child.[54] Seeking to avoid the secrecy and shame that has attended intersex management for a century, intersex activists want doctors to use the same ethical care with these conditions as they would with any other medical concern.[55]

But for most of the twentieth century, many doctors believed that secrecy was the more sensible approach. Dr. Charles Creevy, after finding that a patient who appeared to be a thirteen-year-old girl had testicular tissue, said, "Having determined the sex, it may be wisest merely to inform a responsible relative while leaving the patient in ignorance." His reasoning was that ignorance "avoids the depressing effect of the knowledge of sexual abnormality." Someone in the family should know the truth in case of possible "future difficulties such as a childless marriage or an unexpected change of secondary sex characters and psychosexual outlook later in life."[56] Dr. Emil Novak would have agreed. With regard to the young college woman in whom he found testes, Novak wrote, "There seemed no advantage, and possibly much disadvantage, in having her know the masculine character of the gonads that had been removed . . . The natural reaction would be to look upon their presence in such an individual as a 'contaminating' male factor."[57]

It would not have been unreasonable, according to the standards of the time, for Novak to suggest that his young woman patient become a man. Thus far we have seen instances where doctors encouraged their patients to switch genders based on their gonadal sex. "Certainly, however, it would seem absurd in such a case as this," Novak stated, "to try to adapt the external genitalia for male purposes, even were this technically possible, aside from the tremendous psychic upheaval that such an attempt would have inevitably entailed." The limitations of surgical technique aside, Novak considered his patient's psychological state first rather than relying on the evidence of the gonads alone. "The sociologic factor must be the guiding one in the management of these cases and much less attention should be given to the sex of the gonad than to such considerations as the patient's psychology, and the sex in which she has been brought up. It would be a *cruel* and *unjustifiable procedure* to try to convert into the opposite sex an individual who has lived as a female for many years, merely because she happened to have a male type of gonad."[58] He even recommended postponing a further operation to lengthen her vagina, until she had adequate "chance to recover her physical and mental equilibrium." Her postoperative treatment also focused on her mental health. He recommended that "it should always be in such cases, with the object of reassuring the patient and making her feel that she could lead an essentially normal female life."[59]

An "essentially normal female life" meant marriage—ironically, because, as we have seen, some doctors believed that hermaphrodites should not marry at all. Nonetheless, with or without menstruation or pregnancy, physicians generally defined "success" by their patients' marriageability. As in a Victorian novel, the "happy ending" was matrimony. Dr. Young's reports of his patients and their surgeries inevitably concluded with sentences like this: "Complete success. Patient married and happy."[60] Or this: "Subsequently, a complete urethra was made, and the patient married a charming woman with whom he is living happily."[61] Sometimes, in regard to men, heterosexual intercourse, unsanctioned by wedlock, was highlighted: "The sexual act was frequently indulged in and was entirely satisfactory. He was a happy and entirely contented male."[62]

A patient who did not want to marry or have heterosexual sex raised the troubling specter of homosexuality. Dr. G. Norman Adamson of Chicago saw a patient, an unmarried schoolteacher, whom he called "Miss X," taking care to put her name and feminine pronouns in quotation marks each time he referred to her. He described her as a woman with a soprano voice and pendulous breasts, but hinted that she never played with dolls and preferred active sporting games. During adolescence, "certain complex sexual inclinations began to develop, causing 'her' to be in doubt as to whether 'she' preferred the company of girls or boys." She never menstruated and had not yet had sex with either men or women, despite her attraction toward women. She had an enlarged clitoris and internal testes; Adamson wondered, "The question arises as to the possibility of helping such a patient adjust 'herself' to the social order by surgery."[63]

Should the patient continue living as a woman, and what should her body look like? pondered Adamson. He did not favor an artificial vagina. Considering the high mortality rate of such a procedure (he estimated 6 percent), he called it a "meddlesome surgery" and argued that it would not correct the fundamental problem.[64] But what did Adamson consider the fundamental problem: her intersex condition, or her sexual attraction to women? Did his doubt about creating an artificial vagina derive from his suspicion that "Miss X" was truly a man because of her male gonads?

What did the patient herself want? She said that though she had been raised a girl, she would rather have been a boy. Adamson wrote that she was becoming more and more "dissatisfied with 'her' plight at being far afield in the 'No Man's Land of Sex.'" Perhaps it made more sense to make her into a man; to that end he considered circumcising his female patient's "phallus" in order to make her look more male and excising the undescended, presumably nonfunctioning, and potentially cancerous testicle. Perhaps fashioning a somewhat more typical male body would help "him" adapt to the social order by avoiding homosexuality, given the sexual attraction to women. In the end, Adamson did nothing, concluding

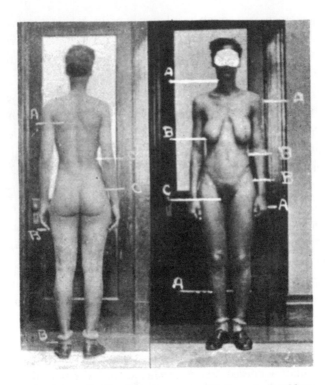

Fig. 1.—(Left) Showing masculinity of shoulders (A) and extremities (B). Note feminine conform'ty of waistline and buttocks (C).

Fig. 2.—Note large size of neck, shou'ders and extremities (A) and the feminine type of breasts, abdomen and pelvis (B). (C) Deformed genitalia.

Dr. Adamson's patient, who had breasts, an enlarged clitoris, and internal testes. To complicate matters, she said she would rather have been a boy and that she was sexually attracted to women. In the end, Adamson did nothing and left her in what he called a "No Man's Land of Sex." G. Norman Adamson, "Hermaphroditism (Report of a Case)," *Clinical Medicine and Surgery* 49 (1933): 146. Courtesy New York Academy of Medicine.

that "surgery has little to offer in aiding such individuals to become adjusted in society."[65]

Antipathy toward homosexuality slanted some doctors' decisions as to their patients' "true sex." One person, Louis/Eloise, a twenty-seven-year-old African American, came to the attention of the Philadelphia municipal court when his girlfriend claimed that he was the father of her newborn child. He denied paternity, claiming that the girlfriend had been pregnant when they met and that there had been no vaginal insemination during intercourse. Two court physicians examined Louis/Eloise and indeed found "pseudo-hermaphroditism and inguino-scrotal hernia." In addition, he was found to be a "low grade moron with very poor comprehension and childish reasoning" and a few months later was admitted to the psychiatric ward of the Philadelphia General Hospital for observation. He was placed in a male ward because of his ambiguous anatomy, even though he entered dressed as a woman.[66]

Two doctors had debated Louis/Eloise's sex at birth, advised raising him female, and suggested amputating the "redundant clitoris." The family had refused surgery. That decision was prescient, as now Louis wanted to live as a man. "In retrospect," his current doctors wrote, "the suggested amputation of the phallus at birth would probably have been disastrous to the patient because of his subsequent male tendencies, psychologically at least, and possibly endocrinologically."[67] They wondered whether or not their patient should be considered a homosexual female, especially since the patient's sister (who called him Eloise) insisted that "she" menstruated and experienced monthly breast tenderness and malaise. The patient believed that the name Louis suited him better and he "denied all menstrual phenomena." He considered himself male, though "anatomically somewhat 'half and half.' "[68] The doctors concurred that "it seemed wisest to convert him into a male." "We must consider the possibility of this individual being psychologically a homosexual female; but by making the patient male, *the perversion socially ceases to exist*. This, of course, is desirable from the standpoint of the community."[69]

Sometimes what the patient wanted and what the doctor wanted were at odds. Throughout the 1920s and 1930s, Hugh Hampton Young and the surgeons at the Johns Hopkins University usually conformed to patients' wishes, even though they did not always believe that the patient was making the correct decision. Their reports often betray a subtext implying that their patients would have been better off following the surgeons' decisions. One of Young's cases had been reared a girl and was in love with a man. She "demanded" amputation of an enlarged clitoris, and when the surgeons complied, they found two testicles. When the patient was notified of the discovery she "strongly asserted that she did not wish to be a man, but intended to marry as a female, because she had a vagina

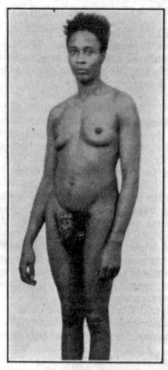

Fig. 1   Fig. 2

Fig. 1.—Picture taken prior to operation showing female contour and large right inguinal hernia and phallus.

    Fig. 2.—Lateral view prior to operation.

This patient's family had refused infant genital surgery and raised Eloise as a girl. Now an adult, he wanted to live as a man, and he expressed sexual interest in women. Worried about homosexuality if Eloise remained female, doctors focused on his male tendencies and stated that if he became male, the "perversion socially ceases to exist." Ralph C. Kell, Robert A. Matthews, and Albert A. Bockman, "True Hermaphroditism: Report of a Confirmed Case," *American Journal of the Medical Sciences* 197 (1939): 828. Courtesy New York Academy of Medicine.

and ordered the surgeon to remove the testicles. As the phallus had already been amputated, this was done." She got married as a woman but later divorced, though not, apparently, because of any difficulties with sexual intercourse, which she said had been fine. Four years later she wanted a lengthening of her vagina, which Young and his team performed. She remarried "and is apparently living happily as a woman, although," Young added, "probably a man."[70]

Throughout this period there was ample evidence that people's sense of themselves as male or female was frequently at odds with what their gonads dictated. Sometimes doctors would try to convince such patients to switch genders; sometimes they deferred to patients' wishes to surgically reconcile their inner and outer apparatus so that they could continue living in the genders in which they were raised. In yet other circumstances, doctors had to content themselves with their patients' desires to live in whatever gender they chose, despite their opposing gonads. Such was the case with a man from Pennsylvania under Dr. James A. Betts's care in 1926. The man was married, though Betts suspected that he and his wife held a carefully guarded secret, as the husband was hesitant to undress for his medical examination. Betts' suspicions were confirmed: the man had broad hips, large breasts, an enlarged clitoris, and, sadly, an ovarian tumor. Betts removed the tumor and found out something of the man's history. Born with ambiguous genitals, this man had been raised a girl, but when twenty-four years old he had decided to live as a man. Betts declared, "I was convinced that my patient was a woman rather than a man," but the man disagreed and continued living as male, tumor and ovary removed.[71]

Some people used their intersex condition to their advantage, displaying their unusual bodies to make money. Francies Benton, a self-termed hermaphrodite, billed himself as "Male and Female in One: One Body— Two People." He exhibited himself very profitably in a sideshow, allowing both men and women—seated separately—to view his body as he disrobed. Dr. Young included a lengthy discussion of Benton in his volume *Genital Abnormalities, Hermaphroditism, and Related Adrenal Diseases*. Benton "had not worried over his condition, did not wish to be changed, and was enjoying life. He was simply curious to know the true sex." Young complied by performing a gynecological examination, but he told Benton that nothing definitive could be said without further exploratory surgery, which Benton ultimately refused. Instead, he requested something from the doctors that would help him: a statement attesting to his hermaphroditic condition "which he might use to convince his audiences that he was telling the truth."[72]

That people might resist surgical invasion should not be surprising.

FIG. 105. Case 15. Photographic copies of advertising material distributed by patient. BUI 24902.

Benton's livelihood was at stake, but for others the physical dangers no less than the social consequences were all too real. During the 1920s and 1930s, doctors were not sure of the course to take once they made their preliminary investigations. As Dr. Walton Martin of New York commented in 1929, "It is interesting to speculate on the wisest procedure in these anomalies which involve not only an unsightly malformation, but are concerned in sex behavior, in the sex psychology of the individual, as well as in legal questions involving inheritance, potency, and legitimacy of offspring."[73] Martin pondered these questions in relation to a deaf, nonspeaking patient who had been raised a girl in an institution for the deaf. When her large clitoris (Martin called it a "penis-like structure") was discovered by a school nurse at age fourteen, school authorities sought Martin's advice. The parents had also noticed the size of the clitoris, but they insisted that their daughter remain a girl. Martin, for his part, feared that her undescended testicles would cause masculinizing tendencies once she went through puberty. Hence his dilemma: "We must decide whether we shall leave the patient alone, to develop as an imperfect male. Shall we attempt to reconstruct the urethra, obliterate the vagina, attempt to make a scrotum and liberate the penis, or shall we amputate the penis and remove the testicles, leaving the child as a supposed girl[?]"[74]

He worried that without surgery, "as sex feeling develops, it cannot be gratified." If doctors left the girl alone, he speculated, she might try to fornicate with the other girls in the dormitory. If they made her into a boy, "he" would have to sit while urinating, and the presence of a vagina "might lead to humiliation and indignities." Reflecting society's generally negative attitude toward the marriage of those with disabilities, Martin was loath to encourage a deaf person to have heterosexual sex. He admitted that if the penis was liberated and the urethra reconstructed, not only would the surgery be difficult and painful, necessitating several operations, but it would "at the best but enable a backward deaf mute to procreate or satisfy the sex instinct." After much indecision, Martin de-

(*Opposite*) Francies Benton billed himself as "Male and Female in One" and displayed his body to paying audiences. These photographs are unusual in that they were commissioned by the subject and do not originate from the medical arena. Hugh Hampton Young, *Genital Abnormalities, Hermaphroditism and Related Adrenal Diseases* (Baltimore: Williams and Wilkins, 1937), 145.

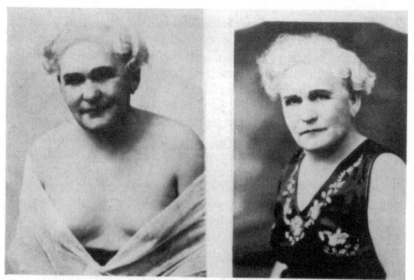

Fig. 104. Case 15. Photographs of patient as a female, showing marked breast development. BUI 24902.

Hugh Hampton Young called Benton a "practicing hermaphrodite, who made a living exhibiting himself-herself in a circus." He described Benton's sexual experiences, emphasizing that he typically had intercourse with females and only had sex with men after his vaginal orifice was enlarged at age 42. Young, *Genital Abnormalities, Hermaphroditism and Related Adrenal Diseases*, 142 (quotation), 144 (picture). Courtesy New York Academy of Medicine.

cided to remove the testicles and amputate the clitoris/penis. He hoped that castration would eliminate the sex drive (thus obviating homosexual sex) and feminize the girl, but he realized that feminization would have been more successful if it had been done prior to puberty. Walton gave thought to the ethics of his decision, but he relied less on the parents' wishes (and not at all on the girl's) and based his decision on his sense that a deaf girl should avoid sexual relations altogether.[75] Ultimately, he argued that other, less disabled children born with well-developed penises, even without testicles, should be raised as boys, "with the hope . . . that the succeeding years will produce the manifesto or evidence of their virilities." In grown children, or adults, he thought it best to "respect the patient's wishes."[76]

Though surgery became the primary mode of medical management throughout the twentieth century, the rush to cut those born or living with sex anomalies was not universal. Dr. Samuel Gross had his critics in the 1840s when he performed surgery on a two-year-old girl with testicles, in a case similar to Dr. Martin's. One hundred years later, in 1941, Dr. Robert A. Ross of Durham, North Carolina, was quoted as saying, "Often the pseudohermaphrodite has been subjected to early damaging surgery simply to confirm a diagnosis, the activating motive more than likely being curiosity rather than surgical judgment. Irreparable harm may be done by such unwise procedures." Ross warned his medical colleagues about an instance of ill-advised surgery for a "psychically derelict" patient. The social service agency involved with the case as well as the police and the parents wanted the patient to become male. The surgery fashioned "fair external genital organs," Ross admitted, but it "also created intense resentment. The subsequent course has been one of continuous lawlessness, meanness, and intractability." From Ross's perspective, "The individual undoubtedly really wanted to be and should have been helped to become a female."[77]

The ethical questions of eliciting the patient's opinion, of informing the patient, and of tinkering surgically were complex. As we shall see, in the 1940s and 1950s, physicians spent a good deal of time assessing patients' psychological makeup before deciding whether to do surgeries that would alter gender presentation. But in the 1920s and 1930s, the interest in patients' psychology was only beginning. Medical opinion held that there was no one right way to manage intersex, in part because the conditions themselves were so varied. Doctors usually involved their patients or the patients' parents in the surgical decision making when the procedures involved only the external genitalia. If, however, they explored internally and found glandular structures opposite to the person's gender performance, then they vacillated about the necessity or wisdom of divulging that knowledge.[78]

We can see why doctors, frustrated by the struggle with patients' and parents' preferences and conflicting indications of sexual composition, ultimately sought ways to manage intersex in infants rather than adults. Babies, doctors came to believe, have not yet had a chance to establish a gender identity, and so surgical alterations would presumably not affect them as it would adults. Parents too, at the baby's birth, could be counseled as to the sex in which to raise their child presurgery. Dr. John Money, the leading researcher in intersexuality for the last half of the twentieth century, came to be the foremost proponent of the theory that an infant's gender was malleable, a theory used to justify genital surgery. By mid-century physicians understood the importance of chromosomes and hormones, but, abandoning the gonadal standard, they elevated external genital morphology as the single most important criterion guiding treatment of hermaphrodites. They firmly believed that, in spite of confounding indicators, social gender could be created to match genital morphology. As Money and his colleagues, Joan and John Hampson, put it in 1955: "For neonatal and very young infant hermaphrodites, our recommendation is that sex be assigned primarily on the basis of the external genitals and *how well they lend themselves to surgical reconstruction* in conformity with assigned sex, due allowance being made for a program of hormonal intervention."[79] This model, assuming the medical assignment of sex and stressing surgical convenience over all other considerations, has had lasting negative consequences. In the next chapter we will see how trends in psychology, surgery, and intersex management in the 1940s and 1950s led physicians to privilege the external sexual organs and to emphasize appropriate rearing as a determinant of gender.

Psychology, John Money, and the Gender of Rearing in the 1940s, 1950s, and 1960s

> In years past the course of action in determining the sex of these patients was decided on the basis of structural and endocrine considerations . . . The evidence indicates that situational and cultural factors played the significant role in the patient's emotional development. Psychiatric and psychological study can define the sexual and social orientation of the patient. Once this is established, the surgeon can transform the external genitalia to fit the psychosexual behavior of the patient.

B Y THE 1940S, doctors were better able to shape people's bodies hormonally and surgically. When ambiguously sexed patients arrived at their offices wanting to have something done to remedy their situations, endocrinologists, urologists, and surgeons turned to drugs and surgical procedures that they hoped would feminize or masculinize their patients. For the most part, physicians responded to patients' requests, as the 1947 epigraph suggests. If a person who had been raised as a girl wanted feminization, doctors complied by amputating her larger-than-average clitoris (which some patients and physicians interpreted as penis-like and embarrassing) and administered estrogen to enhance breast growth and try to stimulate menstruation. Similarly, if a man sought masculinization and chordee (penile curvature) and hypospadic repair, surgeons tried various techniques to bring down undescended testicles, straighten the penis, and move the urinary meatus so that the man could stand to urinate. But many cases were not as straightforward. Most patients' ambiguities extended beyond their external genitalia, and as we saw in the last chapter, doctors labored to ascertain a person's "true sex" by evaluating the individual's gonads and, increasingly, hormonal levels.[1]

However, by the late 1940s, gonads and hormones were no longer thought to dictate a conclusive decision about a patient's sex. Creating congruence between a person's psychology, gender presentation, and external bodily conformation became more important justifications for surgery than matching gonads and genitals. As psychology became more integrated into the medical canon, doctors evaluated patients' attitudes, expectations, and conduct and then molded their bodies to match their patients' sense of themselves as male or female, regardless of gonads. Of course, patients' psychology had always received some scrutiny in intersex medical decision making. When Levi Suydam's gender was questioned back in the 1840s, for example, doctors noticed not only his "feminine" physical features but also his "fondness for gay colors, for pieces of calico, comparing and placing them together," and they decided that his behavior leaned toward the womanly.[2] A century later doctors and psychologists believed that patients' psychological proclivities could be measured more accurately by objective psychological tests than by subjective observation to determine whether, in cases of ambiguity at birth, patients had been reared in the right gender and had adjusted successfully.

By the mid-1950s, when Johns Hopkins University psychologist John Money (along with Joan G. and John Hampson) wrote a series of influential articles about intersex, gender psychology, and genital surgery, they insisted that an intersex person's psychological health depended on the parents' ability to raise an undisputed boy or girl, despite whatever internal incongruities the child's body might evidence. The child psychologist Heino Meyer-Bahlburg later dubbed the hypothesis the "optimum gender of rearing" (OGR) model.[3] Since people, unless informed, were unaware of their chromosomal or hormonal status, what came to matter most to optimum personality integration was the congruity between external genitalia, the sex of rearing, and the patient's psychological sense of well-being as male or female.

The intersex protocols of Money and the Hampsons, as the articles and their recommendations from the mid-1950s have come to be called, profoundly influenced the direction of intersex management. It is not an exaggeration to say that collectively they constituted the essential writing on the subject until the founding of the Intersex Society of North America (ISNA) in 1993, which directly challenged their theories. That Money and the Hampsons' authoritative essays became so respected should not be surprising, given the hundreds of conflicting articles that preceded them. Doctors had disputed the definition of "true sex" for decades. Were the gonads the primary indicators? Were hormones the most important factor? Should intersex people change their sex as adults? Money's essays had definite answers for all of those questions and more. His bold articles were unusual in their conviction (even in comparison with writings by doctors who characteristically conveyed certainty), and they must have come as a welcome relief to professionals confused by the many differing solutions that physicians had put forth. Almost immediately, Money's protocols took hold, and doctors throughout the country believed, erroneously, that the riddle of hermaphroditism had been solved.[4]

Psychology versus the Gonads

The burgeoning emphasis on psychology in relation to rearing was neither unique nor novel. In the 1920s and 1930s, anthropologists and psychologists had developed an interest in what became known as culture-and-personality studies, which explored how personality was related to the demands of the broader culture. The study of child-rearing practices and the emphasis on gender dynamics promulgated by popular anthropological writers such as Margaret Mead and psychiatrists such as Sigmund Freud formed the backdrop for the emergence of psychological theory. At the same time practitioners of both anthropology and psychology sought to make their fields more "scientific," and by midcentury a new tool had entered the medical arsenal: psychiatric testing.[5] We can see in intersex cases throughout this period how theories on the relationship between gender development, individual psychology, and child rearing coalesced. By the time the Money and Hampson team wrote their articles, readers were primed to acknowledge that personality could be nurtured and created, influenced, and forged. In the case of intersex children, Money, Hampson, and Hampson argued, deciding on the gender of the developing personality was the most important factor; then it was up to parents to reinforce the chosen gender to make sure that it "took," to endocrinologists to shape the body with hormones, and to surgeons to craft the correct corresponding genitals.

Many cases published in the 1940s had emphasized the prevailing notions of gender, culture, and personality, on which the 1950s intersex protocols rested. In 1942 the physicians Jacob E. Finesinger, Joe V. Meigs, and Hirsh W. Sulkowitch from Boston detailed what they described as the first case of verified male pseudohermaphroditism to be studied psychoanalytically. As a male pseudohermaphrodite, their patient had female external genitalia (albeit a large clitoris) and some female secondary sex characteristics (though not breasts), but she had internal testes instead of a uterus, ovaries, and other female reproductive organs. She had lived her entire life as a girl, and now that she was seventeen, she wanted to understand why menstruation had never started and why she had not developed breasts. Her obvious lack in this regard "made her feel at times that she could not get married and have children. She could not resign herself to this state of affairs and was in hopes that she could become a normal girl."[6]

Physicians in the 1940s and 1950s described the dreams and ambitions of boys and girls, men and women in stereotypically gendered ways. When they examined the fantasies and anxieties of their patients, they reported that those raised as girls focused on dating, marriage, and children as goals; their reports on boys focused on leadership skills and athletic and financial success.[7] The physicians' insistence on normative gender roles differed little from the gender ideology expressed by doctors in earlier eras, except that medicine now had more sophisticated and "scientific" psychological tools with which to "prove" patients' femininity or masculinity. After exhaustive psychological testing, Finesinger and his coauthors concluded that this patient was an "outgoing, dull person, with a moderately low intelligence." She was slovenly in appearance, though she did use lipstick and nail polish. She liked to socialize, and she thought about boys and dances, but she worried that her lack of menstruation would prevent her from having children. Her fantasies corroborated the physicians' sense of her as female: she identified with her mother and feared her father. She had anxiety dreams about her father's violence (indeed, she thought he had a hand in her mother's premature death) and seemed to transfer these fears to men more generally, believing that "men were cruel, unreliable, and are not to be trusted." "In brief," the doctors noted, "the psychoanalytic material concerning her emotional development and phantasy life was quite typical of that found in females."[8]

The patient's psychological profile suggested to the doctors that she should remain female, but this assessment contradicted her gonadal and endocrine status, which betrayed testes and an absence of female "follicle-stimulating hormones." In addition, her doctors interpreted her hairiness and flat chest as evidence of "preponderately masculine characteristics." Had she seen the doctors just ten years earlier, they might have suggested that she live as a man instead of a woman. But she had been raised female, and they interpreted her psychology as that of a normal, if somewhat boring, young woman. It was obvious to them that other factors played a more significant role than anatomy. "What these other factors are is at present not clear," they disclosed in an unusual admission of uncertainty. They had trouble understanding why or how a person with every physical indication of maleness could model the mind and emotions of a woman, but their patient clearly proved the possibility. "This study would indicate," they concluded, "that broadly speaking the environmental and situational factors (reared as a girl, identification with mother, relationship to father, etc.) in this patient played the predominating role in her psychosexual and emotional development."[9]

Rearing trumped anatomy, it appeared. That was the message Money, Hampson, and Hampson would later expound. It certainly helped explain the situations of many of the patients that physicians saw, even if they did not understand how the factors of femaleness and maleness could contradict one another so blatantly. In prior cases like this, doctors had removed the offending opposite gonads, fearing future malignancy or further masculinization. In this case, though, the doctors chose to leave in the testes, concerned that their removal "might bring about a menopause" and further masculinize her. Instead, they decided to give their patient large doses of estrogen, hoping that it would stimulate breast development, and operate later to create a vagina, presumably so that she could accommodate penetration, should her hopes of marriage ever materialize.[10] The lesson from this case was one that Money, Hampson, and Hampson later elaborated upon: establish the psychology as male or female first and then surgically shape the genitals to match.

Dr. Finesinger, a Boston psychiatrist, saw a similar patient in 1943; in reporting this case he and his coauthor, the surgeon Francis M. Ingersoll, highlighted the efficacy of psychiatric study. A fifteen-year-old girl with a low voice and no breasts or period complained about her clitoris, which was becoming increasingly enlarged, causing her embarrassment when she wore a bathing suit. After a complete physical examination, endocrinological workup, and a laparotomy, which showed both ovarian and testicular tissue, the doctors said that the next step was to send her "to psychiatric service for psychiatric study. *By such a study the true nature of this patient's sex could be determined.*" [11] Not the gonads, not the hormones, not the external genitalia, nothing save psychiatric study would determine her true sex.

It took seven separate admissions to the psychiatric ward (the second admission entailed twenty-eight interview sessions) for the doctors to determine that "she should remain a female." The patient herself told them she felt like a girl and wanted to remain a girl. Her parents concurred; her father stated, "it would break my heart" if she switched genders. He did not think he could countenance her "return from the hospital in boy's clothing." The doctors administered a battery of psychological tests, including the Wechsler-Bellevue Intelligence Test, the Rorschach and thematic apperception tests, the Minnesota Multiphasic Personality Inventory Test, and the Terman-Miles Attitude-Interest Analysis Test, all of which affirmed "a deviation in the direction of femininity." [12]

"A deviation in the direction of femininity" meant that the patient conformed to traditional notions of what it meant to be a middle-class, white, adolescent girl in mid-twentieth-century America. The tests showed her to be immature and passive with a "tendency to avoid unpleasant social situations." She idealized her mother, showed no interest in striving or competing with males, aspired to be a nurse or a secretary, and fantasized about meeting boys, kissing boys, and "of having intercourse with her future husband." Since she displayed "the qualities of neatness, orderliness and compliance," which, according to Ingersoll and Finesinger "are seen as desirable and leading to success," everyone agreed that she should remain a female. [13] A submissiveness, compliance, and passivity reminiscent of the nineteenth-century prescriptive ideal, the "cult of true womanhood," defined femaleness for these doctors. [14]

Though the physicians' conclusion seemed to contradict the advanced anatomical analyses available (the biopsies and hormonal assays that indicated maleness), it was a new analytical instrument that predominated: psychological testing. The doctors were not merely letting the patient decide, as their turn-of-the-century predecessors had worried they were doing in the case reported by Dr. J. Riddle Goffe and discussed in chapter 3. The medical decisions now relied on the science of the mind, not the more traditional science of the gonads or the newer field of endocrinology.[15] In this case, the doctors argued, the patient's "personality and sexual behavior were feminine, yet the structural and endocrine features were essentially masculine."[16]

What should be done when the scientific evidence of psychology contradicted the scientific evidence of the laboratory? "Psychiatric and psychological study can define the sexual and social orientation of the patient," Ingersoll and Finesinger agreed. "Once this is established, the surgeon can transform the external genitalia to fit the psychosexual behavior of the patient."[17] A patient's psychology constituted an entirely new rationale for genital surgery. Whereas in earlier eras, a patient's affect and sense of gender identity influenced doctors' surgical decisions, they relied first on the genitalia and then on the gonads to discover the patient's true sex and acted accordingly. Now that "sexual and social orientation" was measurable through testing, a person's male or female psychology became the new marker of sex, able to rationalize genital surgery even if the newly created genitals contradicted the gonads. Ingersoll and Finesinger's patient's external genitalia looked more masculine than feminine, and she had to resort to wearing a bathing suit with a skirted bottom to conceal the bulge. "Because of the decision to keep the patient a female, the enlarged clitoris was amputated by Dr. Meigs on April 29, 1946, two and a half years after the original operation (to assess the gonadal status)."[18]

Would Ingersoll and Finesinger have allowed their patient to remain female if she had lacked any ovarian tissue? In most of the published cases of the 1940s, the patients presented either had ovotestes or some mixture of gonadal tissue, making it easier for physicians to justify whatever surgical reconstruction they deemed appropriate. Very few cases led to a complete dismissal of the gonads in determining sex. One case, though, stands out in this regard. In an article translated into English and republished in an American journal, G. Cotte, a French physician from the University of Lyon, wrote about his patient, a twenty-two-year-old woman with no vagina who had testicles and no ovarian tissue at all.[19] A dressmaker by

trade, she had been raised a girl and first saw a doctor for a double inguinal hernia when she was eighteen years old. At that time, doctors removed
two small bodies, which were found on histological examination to be
testicles. We do not know if the doctors told her about her gonadal status,

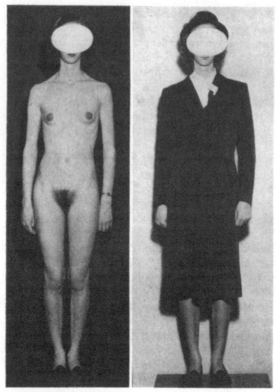

Fig. 376.   Fig. 377.

Fig. 376.—Patient in April, 1947. Note enlarged breasts and darkened areolae
after stilbestrol therapy.
Fig. 377.—Note acceptable female appearance after therapy.

A series of psychological tests revealed that Ingersoll and Finesinger's patient
should remain a woman. These pictures show the efficacy of stilbesterol (estrogen) in promoting breast growth and, as the caption reads, the patient's
"acceptable female appearance" in women's clothing after hormone therapy.
Francis M. Ingersoll and Jacob E. Finesinger, "A Case of Male Pseudohermaphroditism. The Importance of Psychiatry in the Surgery of This Condition," *Surgical Clinics of North America* 47 (1947): 1223. Courtesy New York Academy
of Medicine.

if they advised her to switch genders, or if they just kept her in ignorance. Two years later she sought a doctor's advice because she had not yet menstruated, and she wanted to get married.

Suspecting that his patient would never menstruate, Cotte faced a dilemma as to what course to take. The patient wanted surgical construction of a vagina and threatened suicide if Cotte would not comply. Though Cotte considered her a "male subject," he did not want to "impose" maleness, recognizing the futility of her becoming a man at this stage. Her testicles had already been removed during the so-called hernia operation; her penis was small, and thus she was "incapable of having sexual intercourse as a male." Though she would have physical inadequacies as a male, Cotte also questioned the ethics of enabling her to become more fully functionally female, particularly when he knew her gonadal status had been male. Was he condoning a form of deception? he wondered. Since she had been declared a girl at birth, there was no question that her impending marriage would be legal, but he wondered if surgical procedures perpetuated a charade "by hiding a deformity which involves sterility and which may later be the cause for divorce."[20]

After much deliberation, Cotte was convinced that such intervention was warranted. The woman desperately wanted the surgery, and her fiancé understood her condition and accepted the fact that she would not become pregnant. By inverting her small penis, Cotte created a vagina. Afterward, he noted that the procedure had been successful. His patient married, and sexual intercourse was "easy" and satisfactory to both parties, even though the woman emitted a small amount of fluid (probably prostatic, he suggested) upon orgasm. The surgery, he believed, exemplified what surgeons could and should do for their patients: "modify malformations and thereby ease their effect upon the psyche."[21]

Cotte's patient, if she was viewed as a woman, demonstrated heterosexual desire, which, as we have seen in previous chapters, was a telling consideration for doctors contemplating various surgical procedures for their patients. The threat of homosexuality endured in the 1940s, and doctors still wanted to see their patients happily married rather than, as the psychiatrist Albert Ellis put it, "resorting to homosexuality, bisexuality, or psychosexual immaturity in order to meet the exigencies of an unusual external and internal situation."[22] In 1945 Ellis reviewed eighty-four published cases of hermaphroditism to study "the question of normal and abnormal sexual behavior." Hermaphrodites offered "a beautiful experimental situation all set up for us . . . to observe the sexual psychology they display."[23] His goal was to resolve whether homosexuality was determined by biological (or "physiological") factors or by emotional ones, which he called "psychogenic." Tabulating the libidos of the patients, as far as he could tell from doctors' accounts, he concluded that the overwhelming majority of hermaphroditic patients were heterosexual, based on the gender in which they were raised, rather than on their gonads. The patients who did "resort" to homosexuality, he believed, did so because of their feelings of "social-sexual inferiority," of "being conspicuously 'queer'" because their genitalia did not always conform to their gender assignment.[24]

Easing psychological distress, including that induced by potential homosexual desire, came to be a more important justification for surgery than matching people's genitalia to their gonads. But some doctors speculated that homosexual desire might compel patients to alter their bodies so that they would no longer be considered homosexual. William Perloff and Morris Brody's fifteen-year-old patient, for example, had been raised female and wanted surgery to feminize her body, even though she was found to have internal testes and male chromosomes. Despite psychological testing that indicated feminine interests, "such as dolls, sewing, cooking, and keeping house . . . and an urge to be penetrated," Perloff and Brody thought they might have made the wrong decision in acceding to her wishes for surgery and hormones.[25] They posited that perhaps she was merely "a homosexual male who desired to be a female."[26] They had complied with her wishes, removed the testicles, created a vagina, and administered estradiol [estrogen] pellets to enhance breast growth, but they worried that if the trend toward fulfilling patients' wishes for reconstructive genital surgery was pushed to its logical conclusion, then "genetic males, normal physically, who were homosexual, might present themselves for such treatment. Manifestly, no reputable physician would be a party to such manipulation, and yet, in effect, the present authors were guilty of just such behavior."[27] Their only comfort was that their patient had physical anomalies that created extenuating circumstances, thus somewhat justifying their decision to let her remain female.

As early as 1938, Dr. H. S. Crossen, a gynecologist from St. Louis, exposed the senselessness of relying solely on the gonads for sex determination. He described a patient much like Cotte's and Perloff and Brody's: a twenty-eight-year-old woman with "strong feminine instincts, sexual desires, and general outlook on life, but with testicles and no ovary."[28] In considering constructive surgery, he advised, one needed to know if the patient wanted to live as male or female. In this woman's case, he reasoned, "If I followed the gonads and classified the patient as a male, I was directly opposed by the whole record of the individual's instincts, sexual desires and response and the outlook into the future. If I followed the latter and classified the patient as a female, there was no ovarian tissue on which to rest the decision. A female with testicles and no ovary seemed paradoxical, and of doubtful authenticity." Paradoxical though she seemed, Crossen's patient felt like a woman. He came to the conclusion that patients who had been raised female, "with the right to live as such," despite internal physiological contradictions, should be helped by their physicians. His patient complained of excess facial hair and a short vagina, both of which Crossen alleviated. He administered estrogen, surgically removed her testicles, and stretched her vagina by softening the pelvic tissue.

Crossen wanted to ensure that his patient remained confident of her sex, that is, her femaleness. Anything that might suggest her maleness, he feared, would "disturb her psychologic balance by making her uncertain of her sex." His patient was already distressed by her body's evidence of masculinity. "As a rule," he stated, "she has no thought that she may not be a woman, and it is strongly inadvisable to put her further adrift on the sea of uncertainty by branding her as a male according to the old superficial structural classification."[29]

The physicians J. P. Greenhill and H. E. Schmitz from Chicago echoed Crossen's concerns about the rationale for genital surgery in a paper presented in 1939. Though they recognized that some patients were pleased with surgery that aligned their internal and external conformations, they questioned whether such surgery was always necessary.[30] "Should all individuals who have ovaries and no testicles be made to take their place in society as women regardless of their external genitalia and sexual inclinations and should all persons with testicles and no ovaries be transformed into males no matter what their external genitalia and sexual preferences are like?"[31] Such rigid conformity, they held, "naturally led to much unhappiness and trouble."[32] Greenhill and Schmitz even suggested that there was no need to perform an internal laparotomy to determine a patient's gonadal sex. The point of surgery was not to align gonads with genitalia, but rather to "assist the person to continue to live the sex 'he' prefers or feels inclined to live." Regardless of what gonads were present, they insisted, "If the individual is happily adjusted as a male, 'he' should be helped to continue living as a man and vice versa."[33]

In 1942, Dr. Daniel Chanis saw an eighteen-year-old man whose genitals did not align with his gonads. He had been raised as a boy, had penile erections, and was attracted to girls. At fourteen, he passed blood monthly in his urine. At sixteen, he started having sexual intercourse with a woman who became his common-law wife, and Chanis reported that both found the sex satisfying. His main complaint was his breasts, which could not be hidden under his shirt. At first he appeared to have neither ovaries nor testicles, and no hormonal tests were done because of laboratory technical difficulties. Later, two underdeveloped ovaries were located. Should he become a woman? Chanis wondered. Chanis no doubt considered his patient a woman because of the presence of female gonads. Indeed, he switched pronoun usage in the middle of his report, referring to "her" after this discovery, but he nonetheless decided to aid the patient in continuing his life as male. Chanis believed that psychology (including satisfactory heterosexual sex) took precedence, and so he arranged for breast reduction surgery and a "sub-total hysterectomy." "To change her to her proper social status," he argued, "was too drastic a step to overcome."[34]

By the late 1940s, consensus about the medical management of intersexuality had shifted. Most physicians agreed that surgical decisions should rest on psychological and emotional factors rather than on a strict adherence to the presence of ovaries or testes. Dr. Louis E. Fazen, a surgeon in Minneapolis, unequivocally endorsed what some surgeons had been doing for years: surgical treatment could "bring marked improvement in the psychosexual and emotional characteristics of individuals, even though it is necessary to transform the sex [to the one] opposite to that indicated by their gonads." Fazen went so far as to say that "the anatomic structure of the gonads should be the *least determining factor.*"[35] Dr. Leo F. Bleyer of Madison, Wisconsin, writing in the *American Journal of Surgery,* agreed that reliance on the gonads was an "arbitrary standard." He contended, "The approach from a personalistic psychologic standpoint and the consideration of the total psychophysic pattern seems to be a better way for the practical purpose of a happy adjustment of the patient."[36]

Switching Gender as an Adult

The decline of gonadal significance marked a critical turning point in the medical management of intersex. By increasing the emphasis on psychology and behavior, doctors could turn their attention to the wisdom of intersex adults switching gender. What was required for a successful conversion from male to female, or vice versa? Was the transition even possible? How important was a person's upbringing? Would the switch "take" in all circumstances? Later, in the 1950s, their attention would shift again, away from adults and toward the rearing of children. But in the 1940s, doctors aspired, as one expressed their goal in patronizing terms, "to achieve an acceptable anatomic, physiologic, psychic, and sociological adjustment for these innocently distorted individuals."[37]

In 1944 Grace H. Dicks and A. T. Childers of Ohio published an article that underscored the social problems of gender transition: "The Social Transformation of a Boy Who Had Lived His First Fourteen Years as a Girl: A Case History."[38] Their patient did not have a classic case of hermaphroditism. They suggested that the child was born with male genitals with "a slight degree of hypospadias." The authors' interpretation was that he had been a boy all along, but that the parents raised him as a girl because his mother wanted only daughters.[39]

The patient lived as a girl, Margaret, until the teenage years, when a school nurse uncovered the truth of his male genital anatomy. At that time other male secondary sex characteristics showed their effect as well, and the child's voice deepened. Dicks and Childers did not consider this a typical case of hermaphroditism because, though the child's external anatomy might have appeared ambiguous at birth, his male external anatomy now more or less coincided with his assumed male internal anatomy. According to the writers, "There was nothing in his manner, features, or appearance that would suggest femininity" other than his dressing and living as a girl.[40] Furthermore, the child insisted on living as a boy, named James, once the school physician told him that he was one, and he resented doctors' lying to him throughout the years about his gender status.

A battery of psychological tests confirmed James's masculinity, particularly in the area of mechanical ability. His family concurred that as a girl, he had always been a "tomboy," though sometimes he had done housework and had learned to sew, even though he did not like it. He much preferred repairing the sidewalk with his father. Socially, and apparently physically as well, James was all boy. Even before the school doctor told him he was really male, he had been dreaming for two or three years about life as a boy. "He claimed to have no desire whatever to remain a girl. When it dawned upon him that he really was a boy, he made a definite conscious effort to gain proficiency in all boyish activities . . . He had great confidence in his ability to become a boy at once merely by changing to the proper clothing."[41] The school physician believed that switching genders could be effected easily. All James needed to do was put on boys' clothing and he "could readjust without the slightest difficulty."[42]

Was the shift as simple as changing clothing? Dicks and Childers pondered what it took to switch from female to male. What effect had fourteen years of conditioning had on James? As he had been christened as a girl, would his parents and extended family accept him as a boy? They asked, "Were his interests predominantly feminine or masculine? If masculine, was it because of biological determination and because nature would not be deceived, or had he long realized there was no course but to be a boy in spite of conditioning in the opposite direction?" James's transition was almost effortless. The authors concluded that he had handled his situation exceedingly well, and they were surprised "that he had not shown any more serious emotional disturbances in the face of a problem which might seem insoluble."[43]

Dicks and Childers's case contradicted what Money, Hampson, and Hampson would later turn into dogma: that social conditioning was the key factor in determining stable gender identity. Here was a person who had been raised unquestionably to be female for fourteen years, and yet, when given the chance, he opted to switch genders. This case suggested that having lived through childhood in a particular gender was not enough to create a stable and happy gendered self.[44] In Dicks and Childers's case, their patient's transition was acceptable, even understandable, since he had a penis and had developed male secondary sex characteristics during puberty. His upbringing, he came to believe, had been a mistake, though modern readers might question the doctors' description of how it happened. Was he raised a girl solely because his mother favored daughters? James's hypospadias might account for the mistake, as the degree and severity of the condition might have led a midwife and parents to misidentify a hypospadic penis for female anatomy. In the absence of further hormonal and gonadal testing, it is not too difficult to imagine how James could have grown up as Margaret, only to find out at puberty that his body was developing in the direction of manhood.

To the doctors' amazement, James adjusted beautifully to his new social role. He had been placed temporarily by a child guidance clinic into an institution so that they could monitor his transition. James worried initially about going to the bathroom as well as the "naked parading and easy vulgarity of boys," but he soon realized that his genital condition went unnoticed by the other boys. He was able to become a successful boy, even without hypospadias repair, which he steadfastly refused. The doctors surmised that he had castration anxiety, and that if his hypospadias impeded his ability to live easily among the other boys he would have acceded to it. But James apparently preferred to leave well enough alone, a decision that challenges what John Money later preached: that a boy could not be a boy without a "normal-looking" penis or without being able to stand to urinate. James proved otherwise. He was content with his penis the way it was, and he was so well adjusted socially that there was no need for intensive psychotherapy.[45]

Standardized psychological tests that studied the relationship between sex and personality demonstrated that some intersex people, particularly those who had been raised as girls like James, could become men as teenagers or adults with no adverse consequences. In fact, contrary to what they might have expected, doctors admitted that their patients who switched gender had stable and well-adjusted personalities, with psychological tendencies, interests, and aptitudes that closely matched their new role. Martin Murgy, for example, the pseudonymous patient of the psychologist Catharine Cox Miles of Yale University, had been raised as a girl for seventeen years but had found several aspects of his childhood, such as housekeeping chores, "very irksome." When several physicians "came to the conclusion that he was fundamentally a male," Martin simply cut his hair short, starting wearing male clothing, and changed his name.[46] Surgeons offered Martin a choice of two treatment plans: one that would "eliminate the male, the other the female factors."[47] Though Martin said

TABLE I

STRONG VOCATIONAL INTEREST TEST (1938)

Martin Murgy   January, 1940
Age 20 years.   Sex: Male
Education: High-school graduate

Occupation	Raw score	Standard score	Rating	Occupation	Raw score	Standard score	Rating
Printer..........	+ 59	54	A	School man......	− 7	..	C*
Forest service....	+119	49	A	President manufacturing concern..........	− 17	24	C
Boy Scout master.	+102	..	A*				
Group I.......	+ 64	49	A				
Policeman........	+ 91	46	A	Sales manager....	− 25	23	C
Dentist..........	+ 70	45	A	Carpenter........	− 30	21	C
Musician........	+ 38	45	A	Mathematician...	− 67	18	C
Farmer..........	+ 37	45	A	Advertising......	−109	16	C
Group X.......	+ 10	43	B+	Real-estate salesman..........	−221	14	C
Chemist........	+ 70	42	B+				
Group II......	+ 47	41	B+	Office clerk.......	− 68	13	C
Physician........	+ 38	41	B+	Minister.........	− 67	12	C
Mathematics-science teacher....	+ 42	40	B	Certified public accountant......	− 53	7	C
Y.M.C.A. physical director....	+ 73	39	B	Life-insurance salesman.......	−198	3	C
Author-journalist.	+ 35	39	B	Purchasing agent.	−112	2	C
Group V.......	+ 6	38	B	Accountant......	− 84	− 2	C
Artist..........	+ 59	37	B	Y.M.C.A. general secretary.......	−175	− 4	C
Lawyer..........	+ 27	36	B				
Architect........	+ 34	32	B−	City-school superintendent......	−163	−10	C
Production manager..........	+ 4	32	B−	Vacuum-cleaner salesman.......	−220	..	C*
Engineer.........	+ 33	31	B−				Percentile
Psychologist......	+ 44	30	B−				
Group IX......	− 34	30	B−				
Social-science teacher........	− 3	30	C+	Masculinity-femininity..........	+ 82	52	61
Personnel........	+ 4	27	C+	Studiousness.....	− 48	..	49
Group VIII....	− 22	26	C+	Occupational level	+ 2	48	42
Banker..........	− 27	26	C+	Interest maturity.	− 21	44	17
Physicist........	+ 90	..	C*				

Occupations he has thought of entering: Army, navy, aviation, technical work in movies, exploring.

* From 1933 Strong.

Catharine Cox Miles's patient, twenty-year-old Martin Murgy, scored high interest levels in traditionally masculine jobs, such as the military, police work, the forest service, and Boy Scout master, even though he had lived as a girl for seventeen years. Catharine Cox Miles, "Psychological Study of a Young Adult Male Pseudohermaphrodite Reared as a Female," in *Studies in Personality, Contributed in Honor of Lewis M. Terman*, ed. Quinn McNemar and Maud Merrill (New York: McGraw-Hill, 1942), 216.

he could live as either a man or a woman, he chose to become a man because of the independence and freedom manhood would offer. The standardized tests he completed confirmed his decision. Though he had no
"special drive or exceptional ambition," Miles noted, Martin seemed, "on
the whole, to be one of those persons of average composure and balance
who have practically no psychopathic difficulties and no serious personality conflicts. In view of his peculiar physiological development and life
experience," Miles remarked, "this seemed rather remarkable."[48]

Martin took the Stanford-Binet Intelligence Test, the Otis Speed Test
of Intelligence, the Otis Higher Intelligence Examination, the Terman-
Miles Attitude-Interest Analysis Test (which Miles helped develop), and
the New Stanford Achievement Test, among others. Collectively, the tests
showed a rather "typical male adolescent profile" (though Martin avoided
mathematics), and interests in classically masculine occupations, such as
aviator, explorer, foreign correspondent, and forest ranger. To the extent
that his psychological tests supported mixed results, the researchers had
a rationalization: Martin's tendency toward masculinity could be explained
by his single testicle; any tendency toward things feminine could be explained by his upbringing as a girl.[49]

Standardized psychological tests, or psychometry, could accurately
evaluate personalities, proponents believed, especially behavioral differences between men and women, because subjects could be analyzed in
relation to others.[50] Martin's test scores were compared with those of
heterosexual men and women of his age group and educational level, as
well as with those of male and female "inverts." Catharine Cox Miles had
been a graduate student of the psychologist Lewis M. Terman at Stanford
University. Together they developed a model that was believed to determine the relative masculinity or femininity of their research subjects. Most
men, they found, had characteristics that could be called masculine, and
most women had what could be called feminine traits, behaviors, and
interests. In their 1936 book, *Sex and Personality: Studies in Masculinity
and Femininity,* they outlined the M-F Test, a seven-part diagnostic tool
that became popular among not only doctors and psychologists, but employers, researchers, and school districts as well. The point of the test was
to standardize male and female behavior, just as Terman had done earlier
with intelligence testing in the Stanford-Binet Test. Test takers were not
supposed to be aware of the purpose of the test or the significance of their

TABLE II
M-F Scores of Individuals and Groups

Individual or group	Number	Exercises 1	2	3	4	5	6	7	Total	Standard core
Martin Murgy:										
Form A..........	1	− 8.0	+1.0	+ 2.0	+14.0	+ 58.0	−5.0	0	+ 62.0	+ .45
Average, A + B..........	1	− 8.5	0	− 2.5	+15.0	+ 90.0	+ .5	−1.3	+ 93.4	+1.05
Form B..........	1	− 9.0	−1.0	− 7.0	+16.0	+122.0	+6.0	−2.7	+124.7	+1.60
Adults, 20 years old, group average, high-school education:										
Matched group A..........	26	− 5.2	0	+ 6.6	+18.4	+ 41.1	+1.5	− .4	+ 62.0	+ .45
Matched group B..........	16	− .1	+ .6	+ 6.5	+35.3	+ 81.0	0	+1.1	+124.5	+1.60
Male-norm group averages:										
Adults, 20 years old (I)..........	35	− 4.0	+ .2	+ 6.6	+24.9	+ 53.9	+1.2	0	+ 82.9	+ .85
High-school education (II)..........	24	− 6.8	0	+ 5.8	+20.2	+ 39.5	+1.4	− .4	+ 59.6	+ .40
Adults, total population*..........	1083	+ 2.8	0	+ 2.2	+21.5	+ 23.0	0	− .8	+ 43.3	+ .10
High-school boys*..........	308	− .8	0	+ 5.2	+14.4	+ 56.7	+ .8	− .7	+ 77.1	+ .75
College men*										
Small stature..........	43	− 8.2	0	+ 8.5	+27.8	+ 29.6	+4.3	−1.3	+ 61.7	+ .45
Special male group averages:										
Inverts*..........	77	− 5.7	− .9	− .1	+21.3	− 36.2	−2.8	−2.7	− 28.0	−1.20
Inverts†..........	26	− 8.1	− .6	− 3.1	+30.7	− 21.3	− .7	−1.3	− 4.4	− .75
Effeminate men†..........	13	−10.5	− .7	− 7.7	+27.1	− 23.4	−3.4	−2.4	− 21.2	−1.60
Female norm group averages:										
Adults, 20 years old*..........	604	−14.2	−1.0	− 8.5	+ 4.1	− 45.5	−5.6	−2.7	− 74.2	+ .25
Adults, total population*..........	1867	−14.2	−1.0	− 8.9	+ 1.3	− 48.0	−5.8	−2.3	− 80.9	+ .10
High-school girls*..........	245	−14.3	−1.1	−10.7	−12.9	− 32.3	−6.6	−2.4	− 79.3	+ .10
Special female group averages:										
Inverts*..........	18	−18.2	+ .3	− 7.5	+11.5	− 18.0	−1.6	−1.6	− 36.4	+1.15
Inverts†..........	21	−11.7	+ .6	−11.0	+14.2	− 24.1	− .4	− .9	− 34.6	+1.15

* Data from Terman and Miles (9), Tables 96 and 97.
† The scores from Henry's (2) published data for Exercises 2, 5, and 7 are given here, recalculated to agree with the other figures in this table.

Patients such as Martin Murgy underwent various personality tests that ranked their masculinity or femininity relative to other heterosexual men and women in their age group. In this test, Murgy's scores could be compared as well to male and female "inverts" and effeminate men. Catharine Cox Miles, "Psychological Study of a Young Adult Male Pseudohermaphrodite Reared as a Female," in *Studies in Personality, Contributed in Honor of Lewis M. Terman*, ed. Quinn McNemar and Maud Merrill (New York: McGraw-Hill, 1942), 218.

answers. In other words, they were not supposed to be able to figure out the "right" responses.[51]

The M-F Test, along with the Sex Variant Study (a more extensive survey developed in the mid-1930s that involved physical measurements as well as psychological profiles), was supposedly able to distinguish "normal" heterosexual men and women from homosexuals. The test could indicate when something had gone wrong, in other words, when a person had failed to acquire gender-appropriate identity and thus had become an invert, or homosexual. Based on Martin's answers in the M-F Test, Miles was not sure whether Martin qualified as an invert. In some respects he scored similarly to the male inverts; in others his scores seemed more like those of the female inverts, particularly those who had "to conform to feminine social customs while protesting inwardly."[52] Martin's emotional health militated, however, against a diagnosis of invert. "Inverts are known to have many psychopathic traits," Miles explained, yet Martin was a "normally stable, well-adjusted person."[53] Finally, Miles became convinced that "relieved of his anatomical anomalies and of the psychological restrictions of enforced femininity, he was quite ready to be a man in attitudes and functions."[54]

Failure to acquire an appropriate sense of oneself as male or female was standard for an intersex person, some doctors believed. In 1948, Dr. Rita Finkler, a pioneering endocrinologist in New Jersey, stated unequivocally that "the intersex individual has always been a problem to himself and to his social group" because at some point the patient recognizes the incongruity between the sex of rearing and the genitalia or secondary sex characteristics.[55] Finkler's patient, W. D., had been raised as a girl but at age nineteen developed "symptoms of anxiety neurosis" when she became sexually attracted to other girls at the same time that she became aware of an enlarged clitoris. She experienced "palpitations, dizzy spells, tinnitus, sudden blackouts, headaches, anorexia and insomnia." Since she wanted to be a man, doctors did bioassays and psychological tests, all of which confirmed that she was more male than female. Doctors pronounced her an "unfinished male" and sought to align her hormonal and physiological states with her desire and penchant for masculinity. Intensive testosterone therapy ensued, as well as a surgical procedure to free what they interpreted as a bound-down penis. After several months of testosterone therapy, W. D.'s voice deepened, his genitals and muscles developed, and he decided to change his legal status from female to male.[56] As he had always worn masculine-looking clothes and short hair, the transformation was not abrupt, and he "expressed great relief and satisfaction in the change so ardently wished for in the previous two years."[57]

Many physicians were surprised at the ease with which their patients could switch genders. Some had assumed, as Dr. Leona M. Bayer of Stanford University stated in 1947, that it was too difficult to "change horses in mid stream," or to "change sex in midlife."[58] The cases of successful switching could suggest that something biological was at work; that something in the body was guiding an ambiguously sexed person toward one gender or the other. These patients raised a still-unanswered question: was one's sense of gender identity based on an innate biological basis, or did it depend on social and environmental factors, such as education and upbringing? Which force was decisive—nature or nurture?

Though Dr. Bayer wrote that there was no proof "that biologic bases for masculinity-femininity attitudes are *not* present," she admitted uncertainty. It was "difficult," she said, "to assign either biological or environmental causality to the psychological trends." She concluded, "It would therefore seem desirable to make determinations of the basic sex as early as possible and to institute guidance before the child has laid down permanent patterns of behavior."[59]

Other doctors concurred that deciding on a sex at the earliest age possible was optimal, in the hopes that the correct decision would preclude switching later in life. The physicians Charles Morgan McKenna and Joseph H. Kiefer, urologists from Chicago, urged in 1944 that changing the sex of adults be avoided. "When such a change is necessary," they said, "the earlier in life it is done, the easier for the patient, his family and all concerned."[60] McKenna and Kiefer reflected on the outcomes of Hugh Hampton Young's patients from the 1920s and 1930s and asserted that "children will usually accommodate themselves to the sex in which they are raised." Of the twenty cases of true hermaphroditism (people with both ovarian and testicular tissue) on which Young reported, only one had strayed from the sex of rearing. These patients could have turned out to be men or women, McKenna and Kiefer implied, because their gonadal status was mixed. Yet overwhelmingly they chose the gender in which they had been raised whether because their parents or birth attendants had chosen correctly or because changing genders later was too complicated and embarrassing. "After school age the change is more difficult," McKenna and Kiefer argued, "and is accompanied by psychic trauma and social uproar."[61]

Of course, early decision meant a reliance on gonads and hormones, as opposed to a person's psychology, which could not be known in infancy. Even though the trend throughout the 1940s had been to focus on a person's emotional life, rather than gonadal status, it was impossible to examine that realm when the patient was a baby. McKenna and Kiefer recognized that some patients' gonads contradicted the "body characteristics, psychic orientation and libido" manifested when they got older, yet they nonetheless maintained, "we are sure that fixing these individuals in the most suitable sex early in life offers them the best chance of avoiding difficulties later."[62] McKenna and Kiefer were among the earliest proponents of what Money and the Hampsons later advocated. At first glance, theirs seems a reasonable proposition. If most people stick with the gender in which they were reared, it makes sense to ensure that the right gender has been chosen in infancy. But, as we have seen, there was no sure way to choose the gender of an infant born with ambiguous genitalia, particularly an infant who also had mixed gonadal status and hormone levels.

The Malleability of Gender

John Money and his colleagues had a rationale for choosing gender that came to define intersex management for the next fifty years. Though physicians understood the importance of chromosomes and hormones, most followed Money's advice and elevated external genital morphology as the single most important criterion in deciding how to treat intersex people, firmly believing that, in spite of confounding indicators, social gender could be created to match genital shape. In 1957, Money and the Hampsons wrote in one of their influential articles: "The chromosomal sex should not be the ultimate criterion, nor should the gonadal sex. By contrast, a great deal of emphasis should be placed on the morphology of the external genitals and *the ease with which these organs can be surgically reconstructed* to be consistent with the assigned sex."[63]

Money believed, speciously by today's appraisal, that one's sense of gender identity was malleable until about eighteen months of age. He therefore concluded that those born with ambiguous genitalia could have their gender surgically assigned as infants (and later reinforced during puberty with hormones) without negative consequences. Once their bodies were surgically formed to approximate male or female models (usually female because it was easier for the surgeons to create female-style external genitalia from ambiguous ones), the children would develop personalities matched to their assigned gender, provided the assignments were supported by proper rearing via parental commitment to the chosen gender.

How John Money came to offer these influential recommendations requires some explanation and analysis. In some ways they seem to contradict a central tenet of his 1952 Harvard dissertation, which postulated that intersex people *did not* suffer from neurotic symptoms or psychoses. Money had argued in his dissertation that, despite what common sense might suggest, people with ambiguous genitalia did surprisingly well psychologically. "Even the stringent emotional problems of living an inexorable sexual paradox," he proclaimed, "do not necessarily entail psychiatric disorder."[64] Time and time again, Money evaluated the subjects he profiled in exceedingly positive terms. Of one patient, he wrote, "He is meeting life most successfully without any suspicion of psychopathology. There is every reason to believe he will continue to do so. His life is an eloquent and incisive testimony to the stamina of human personality."[65]

If intersex patients did so well psychologically, despite incongruous genitalia, then why bother with infant genital surgery at all? The answer lies in Money's commitment to gender-appropriate rearing, a view that he set forth in his dissertation and propounded in subsequent articles. Money believed that parents would be better able to raise an intersex child in the suitable chosen gender if that child's genitals looked as "normal" as possible.

It stood to reason, he argued, that children would want their genitals to match the gender in which they were being raised. And the sooner this could be done in their young lives, the better off they would be. If parents waited until their children were older for such operative procedures, the children might have to endure years of teasing and would develop insecurities about their bodies.[66] If the surgery were done early in infancy, Money concluded, patients would have their entire lives to adjust more easily to the gender deemed most suitable. The decision of which gender to choose would be based, according to Money's philosophy, not on the gonads, hormones, chromosomes, or psychology, but rather on the ease with which the genitals could be surgically shaped.

As we have seen in the past two chapters, doctors had not agreed on any single criterion on which to base gender assignment (even the intersex person's own assessment of his or her gender was not sufficient for some). In a 1953 article, the authors noted "the problem of hermaphroditism appears to have reached an impasse and a new approach is desirable."[67] Lamenting the confusion, Dr. Charles Hooks wrote in 1949 that physicians faced "an unenviable enigma," making clinical management "more arduous and harassing."[68] Ironically, the disagreement among physicians may have encouraged the quick and eager acceptance of Money's assertions. It must have seemed that someone was finally offering concrete and authoritative answers to many vexing questions. As Suzanne Kessler, a professor of psychology and author of the 1998 book *Lessons from the Intersexed,* has noted, his theories corroborated "contemporary ideas about gender, children, psychology, and medicine. Gender and children are malleable; psychology and medicine are the tools to transform them."[69] Money's confidence in treatment dictated by the morphology of the genitals seems, at least in hindsight, overly simplistic. Though physicians had disagreed for over a century about which factors were the most important in determining sex, nearly everyone agreed that several needed to be considered.[70] The gonads, long considered the gold standard, had waned in significance, but other criteria, particularly hormones and chromosomes, and, for older patients, psychology, all needed assessment. Money did not completely abandon those other criteria, but in each publication he emphasized that the ability to craft genitals that most closely approximated those appropriate for boys or girls would best ensure steadfast rearing in that particular gender and, hence, psychologically healthier patients.[71]

Joan G. Hampson, a psychiatrist and John Money's colleague at Johns Hopkins, conceded in a 1955 article that the use of chromosomes, gonads, and hormones to determine "psychosexual orientation" had long "been . . . favored by the prevailing climate of medical opinion." According to Money's clinical evidence, she argued, there was nothing to support the assumption that one's gender role was automatically determined by those factors. "On the contrary," she said, "the evidence indicated that a person's gender role and awareness is founded in what, from infancy onward, he learns, assimilates and interprets about his sexual status from his parents, siblings, playmates and others, and from the way he reads the signs of his own body."[72]

Money and the Hampsons' understanding of the intersex person's gender orientation paralleled broader trends in psychology throughout the 1950s. Like other psychologists and psychiatrists who published then, they asked questions and observed their patients' "general mannerisms, deportment and demeanor; play preferences and recreational interests; spontaneous topics of talk in unprompted conversation and casual comment; content of dreams, daydreams and fantasies, replies to oblique inquiries and projective tests; evidence of erotic practices; and finally the individual's own replies to direct inquiry."[73] What they found (and what Money had observed in his dissertation) was that most people with ambiguous genitalia remained content in the gender in which they were raised.[74]

Money had insisted in his dissertation that what children learned from each other, their parents, and their own bodies was critical in forming what he called their "gender role." Rearing children unambiguously came to be the centerpiece of Money and the Hampsons' guidelines. The genitals had to look "normal," they said, or parents would not be able to do their job of rearing effectively. If parents looked down at their baby girl's genitals and saw a large clitoris that looked more like a penis, it would confuse them and cause them to treat their child more like a boy than the girl she was supposed to be. This approach, which centered on rearing based on the physical evidence of (surgically altered) matching genitalia, had far-reaching, and, as we shall see, often negative consequences.

Money's earlier emphasis on his adult patients' psychological well-being had the potential to transform intersex management beneficially. If doctors and psychologists had continued to listen to their patients, as they had throughout the 1940s, and surgically transformed bodies only as the patients deemed necessary, the surgery would have been ethically justifiable. Instead, the Hopkins team focused their attention on infants. How, they asked, can we ensure that intersex infants and children will grow up psychologically well balanced? Money had encountered intersex adults who had suffered embarrassment because of their unusual genitals, but even those people, he argued, had overcome it. Yet although Money stated in his dissertation that intersex people could grow up psychologically healthy, he came to advocate "correcting" as much as possible in infancy, thereby giving the child the best possible chance to be reared unambiguously.[75]

Money's dissertation findings might have led him to conclude that since intersex people generally grow up content living as the gender they were raised, surgery is not necessary, unless intersex people seek it for themselves as adults. But instead he made a compelling case that since most intersex people remain the gender in which they were reared, it makes sense to ensure their rearing is unequivocal by fashioning genitals that make it obvious to parents that the child is unambiguously male or female.

The Significance of Sex of Rearing

The Hopkins team, including the Hampsons, Money, endocrinologists such as Lawson Wilkins, and pediatric surgeons, effectively changed the nature and tone of the debate concerning intersex management. Their goal was to eliminate doubt and uncertainty for children growing up with intersex conditions. And so in each of the four articles they published together in 1955–56, they sought to understand the factors that contributed to an intersex person's sense of insecurity about his or her gender identity.

Steps had to be taken to avoid "adult equivocation and doubt," as Money's team believed that children would detect any ambivalence and that this recognition would undermine their sense of self. Doubt could arise not just among parents, but among physicians as well. In one article published in 1955, Joan Hampson mentioned troubling cases where the doctors counseled parents to switch the gender of their children because they found conflicting gonads or chromosomes. She praised parents who balked at doctors' suggestions that children switch genders and who relied instead on "their homely wisdom to ignore such recommendation." Hampson's stance can be seen as radical: she advised parents to disregard their doctors and to rely on instinct and what they felt to be true about their children's gender. Hampson was calling for an almost unprecedented confidence in patients' and parents' judgments. Despite Hampson's praise for parents' instincts, the protocols sought to influence parents through the surgically created clues provided by their children's appearance. Of course, the goal was the patient's well-being, but the method was to shape parents and children. Children could have opinions too, Hampson argued, though "it has been rare that they have been given any opportunity to express it."[76] She went so far as to say, "Historically, a change of sex has been imposed more often than consented to."[77] Ironically, by changing the shape of intersex infants' genitals, the Hopkins team itself imposed sex changes as well, since infants had no opportunity to express their opinion or give their consent to surgical procedures that would change their gender.

The Money and Hampson protocols ostensibly sought to avoid a change of sex, though their emphasis on sculpting genitalia made surgical sex reassignment (at least in infancy) common. If a change was to be made, they were certain, it was best done early, before the child became aware of gender. Assigning a gender at infancy depended primarily on the external genitalia. Joan Hampson wrote in 1955, "If the external genitalia cannot possibly under any conceivable circumstances be surgically reconstructed for functioning in one sex, then the other sex should be assigned and subsequent medical and surgical efforts should be directed towards securing hormonal as well as morphological congruity with the assigned sex."[78] A gender change was advisable if the parents could make a smooth adjustment, "tak[ing] their infant's change of sex sensibly and in their stride." Recognizing the difficulty of such "an extremely disturbing experience," Hampson argued nonetheless that changes made in the first year "proved thoroughly successful." She advised against making any such changes in a child more than two and a half years old, when "gender role is indelibly established," and more important, when the parents' "ingrained conviction" of their child as either boy or girl could not be overturned.[79]

The specific guidelines that the Money and Hampson team published a month later in 1955 reiterated the justification for and optimal timing of interventionist infant surgery. "It should be the aim of the obstetrician and pediatrician to settle the sex of an hermaphroditic baby, once and for all, within the first few weeks of life, before establishment of a gender role gets far advanced," they urged.[80] Since genital morphology was the most important determinant, they recommended surgery only for babies whose genitals were ambiguously formed. They recognized that "some hermaphrodites are born with external genitals that look so completely masculine or feminine that ambisexuality is not suspected. The baby is unhesitat-ingly declared male or female and no question is raised perhaps for years." Later in life they might opt, however, for hormonal treatment to correct whatever hormonal anomalies appeared. Beginning in the mid-1950s and 1960s, doctors throughout the country followed Money's advice when "early uncertainty is aroused" and routinely changed children's ambigu-ous genitals to match usual male or female organs more closely.[81]

Money and his team wanted to guarantee that as those infants grew into adulthood they would avoid psychological problems. Like earlier practitioners, they hoped each patient would form a solid gender identity and have a successful sex life, which they assumed would be heterosexual. From our twenty-first-century perspective, it is difficult to credit their belief that certain genital surgeries would not impair sexual function; clitoral removal or even clitoral reduction, for example, removes sensitive nerve tissue and thus thwarts orgasmic response.[82] Yet that is precisely what Money and the Hampsons declared. They stated, "Clitoral amputa-tion in patients living as girls does not, so far as our evidence goes, destroy erotic sensitivity and responsiveness, provided the vagina is well devel-oped."[83] They surveyed a dozen women about their sexual sensation fol-lowing the surgery, and none of them reported a loss of orgasm. In fact, all twelve "were unanimous in expressing intense satisfaction at having a feminine genital morphology after the operation."[84]

It is important to keep in mind that the twelve women were already adults, had already established themselves as women, and had requested a procedure that they hoped would feminize them. The same procedure performed on a baby might have a less positive effect, including the for-mation of scar tissue, which could lessen future sexual response. But even so, the team reassured readers, "If clitoridectomy is performed in early infancy, the chances of undesirable psychologic sequelae are negligible."[85] They did, however, advise that clitoridectomy was not for everyone. If the patient had decided on a masculine gender role, then this surgery would be "most unwise," presumably because a person who considered himself male would want to use the enlarged clitoris (or penis) for penetration.[86]

The assumption that intersex people would be heterosexual went largely unexamined. In a discussion of the twelve women who underwent clitoridectomies, Money and the Hampsons broached the topic of vaginal reconstruction, with the conjecture that the women would want penile penetration. If the surgery had to be delayed until after adolescence, they contended, there was no need to wait until after the patient was engaged or married. She should be free to request the procedure at any time, thereby avoiding the embarrassment of having "to explain and confess her vaginal shortcomings to a prospective husband."[87] They consistently linked femininity with heterosexuality and motherhood: "Subsequently surgical feminization and hormonal regulation can be thoroughly successful, thereby preserving areas of erotic sensation, fortifying the feminine role and safeguarding the opportunity for eventual child bearing."[88]

To the extent that Money and the Hampsons discussed homosexuality, their attitude paralleled the mid-1950s supposition that same-sex desire was a psychopathology best avoided.[89] Money and the Hampsons contended that a gender change at an older age was more likely to bring about homosexuality. If, for example, a boy who was found to have ovaries was persuaded to become a girl, he might "continue to think, act and dream as the boy he was brought up to be, eventually falling in love as a boy, only to be considered homosexual and maladjusted by society." Or, the gender change might very well be effected only partly, "at the cost of psychologic disorder and symptomatology sufficiently disabling to prevent marriage." If parents followed the Hopkins guidelines, however, all of this could be circumvented. Parents needed to be reassured, the team insisted, that "their child is not destined to grow up with abnormal and perverse sexual desires, for they get hermaphroditism and homosexuality hopelessly confused."[90]

Just as gender could be molded and shaped under proper parental guidance, Money and the Hampsons suggested that sexual orientation could be similarly guided. At birth, the sex drive was neither male nor female, they said, and it has "no other somatic anchorage than in the erotically sensitive areas of the body." "In the course of growing up," they suggested, "a person's sexual organ sensations become associated with a gender role and orientation as male or female which becomes established through innumerable experiences encountered and transacted." They posited that years of sexual experience (with heterosexual partners) as a man or a woman would establish genital arousal with the appropriate partner. "From the sum total of hermaphroditic evidence, the conclusion that emerges is that sexual behavior and orientation as male or female does not have an innate instinctive basis," they stated. Without explicitly arguing that either heterosexuality or homosexuality was entirely environmental, they hinted that sexual desire was linked to a potentially impressionable gender identity; most people, including intersex people, could be conditioned to be either women or men with suitable, and psychologically healthy, heterosexual desire.[91]

According to Money and the Hampsons, the healthiest patients were likely to be those whose "body morphology, irrespective of gonads and chromosomes, [was] unambiguous looking than among those whose sexual appearance was equivocal." Those few intersex patients deemed psychologically unhealthy by Money and the Hampsons had, they thought, grown up with doubts about their authenticity as males or females. Though 95 percent of the ninety-four patients they observed in one study could be classified as healthy, the 5 percent who experienced trauma had endured "ridicule, insinuation or misgivings of other people as well as the about-face of a reassignment of sex" *after* early infancy. Their data, they believed, quite clearly supported the conclusion that the earlier genital contradictions were "corrected," the better off the child would be psychologically. Older children or teens who had experienced the same surgical procedures after infancy did not "have their quandary automatically resolved in its entirety once surgical or hormonal corrections are effected."[92]

Money and the Hampsons believed that changing one's gender later in life was too difficult a transition to navigate effectively, particularly if the change was predicated on gonadal status: "When, in deference to the presumed importance of gonads, a change of assigned sex was imposed later than early infancy, the life adjustment was not significantly improved and was often made worse."[93] They argued that in cases of ambiguity, external structures should guide the decision of what gender a child should be. Once that was decided, even if the person found out that his/her gonads might have dictated the opposite sex, the influence of rearing usually prevailed. They were convinced that most people remained the gender in which they were reared and that only a very few would want to change gender or be able to do so successfully.

There were a few exceptions. "In the exceptional case of an hermaphroditic child who has secretly half-resolved on a change of sex, successful negotiation of a change may prove possible," they admitted. "But our experience has led us to believe that voluntary requests for change of sex in hermaphrodites belong to the teenage. Though such requests are rare, they deserve serious evaluation, for they are usually a culminating attempt to resolve years of well founded perplexity and doubt."[94] Of the seventy-six patients Money and the Hampsons studied for inclusion in one of their 1955 articles, "An Examination of Some Basic Sexual Concepts: The Evidence of Human Hermaphroditism," twenty-three had not had any infant surgery to resolve ambiguity and so had thus lived with a contradiction between their external genital morphology and their assigned sex. With the exception of one person, all had come to terms with the incongruity. Significantly, in all of the seventy-six cases, the authors noted, the patients were psychologically healthy. Though some (and not all) faced "feelings of bashfulness, shame and oddity," psychotic symptoms were "conspicuous by their absence." Their lives had not been easy, the authors granted. There was much "distress and anguish" among some of them, particularly those who had been raised with confusion about their gender. The underlying assumption that Money and the Hampsons made was that intersex teenagers' and adults' years of "well founded perplexity and doubt" were due primarily to their parents' ambivalence about what gender they really were. Had they been raised with more certainty, the authors asserted, they would have been content and would not have felt the need to change anything.

Only one of the seventy-six studied chose to live as a man after he had been raised female. All of the others, the authors reported, were psychologically helped when they chose corrective plastic surgery that reinforced the gender in which they had been raised. Money argued that generally a person's decision to change gender would not be "psychologically pervasive—the person acts his new role, maybe as convincingly as an actor who becomes the character he portrays, but is always capable of leaving the stage to become his authentic self."[95] As if to imply that the one outlier, the one person who chose to become a man, was not "really" certain of his decision, Money, Hampson, and Hampson denigrated his refusal to "take advantage" of plastic surgery. Despite choosing to live as male when he turned sixteen, "He was *unable to summon up enough courage* to have his genitals masculinized."[96]

Notably, though Money believed that "reassignment of the sex of rearing after the early months of life was, without doubt, psychologically injurious" for intersex patients, he did not oppose sex reassignment surgery for transsexuals, those without an apparent intersex condition who sought to change their gender.[97] In the 1950s, when the protocols were established, sex reassignment surgery (SRS) for transsexual adults was in its infancy.[98] In 1952 Christine Jorgensen's transition had made a huge splash in American media, but even a decade later not a single major medical center endorsed such surgery. John Money was an early advocate. Because of his research on intersex patients, transsexuals had sought his advice, and over the years he tried to persuade his colleagues at Johns Hopkins to perform sex-reassignment surgeries. According to the historian Joanne Meyerowitz, an influx of funding, the success of organ transplants, and the more liberal environment of the mid-1960s, particularly on matters of sex and sexuality, coalesced, and by 1966 Money and his colleagues had established the Gender Identity Clinic for transsexuals at Hopkins.[99]

Over the years scholars, activists, and physicians have criticized the ethics of Money's intersex protocols because, unlike transsexuals who chose their own genital surgery (and, in fact, had to convince doctors to perform the procedures), infants could not choose for themselves. Parents, not the patients, made the life-altering decisions about surgery based on doctors' and psychologists' advice for handling intersex infants. Since the 1990s, intersex people have been speaking out publicly about what was done to them as children; some whose genitals were altered in infancy never felt at home in their assigned sex. Some never knew their medical history, as doctors encouraged parents and relatives to keep the matter secret. When they found out, some changed gender as adults. Others struggled to accommodate life with severely compromised surgically altered sexual organs as well as with a deep sense of shame induced by enforced secrecy.[100]

On Secrecy and Concealment

Money and the Hampsons' protocols offered contradictory advice concerning concealment and transparency. On the one hand, they insisted on parental understanding of the child's condition. Most people had only a vague and misinformed notion of what hermaphroditism entailed, they believed, and used such "emotion-laden and embarrassing expressions as 'morphodite' and 'not being right down there.'" The team urged parents to learn the correct medical vocabulary so that they could have more informed discussions with their child. Rather than having parents mistakenly think of their child as "half boy, half girl," they preferred the more "enlightening concept of genital unfinishedness." One article included a simplified chart depicting sexual differentiation in the human fetus that suggested an incomplete path of sexual differentiation in embryonic development.[101]

Parents should understand their child's condition, but no one else had to know, particularly if the child underwent infant genital surgery that effectively decided his or her gender. The protocols encouraged secrecy by advising parents to leave town and begin life anew among neighbors and friends who would never know that the child began life as a boy and had since become a girl. Anticipating difficulties as a child matured, Joan Hampson suggested severing ties with anyone who knew the child before the switch. "Without the move," she warned, "the child is too likely to be harried and disturbed by rumor and gossip in later years."[102] Money and the Hampsons warned, "Later in life the child is likely to be confronted with coarse jokes and reminders, unless the family had started life entirely afresh in a new community at the time of the change."[103]

It seems almost inconceivable that parents who endured the worry and anxiety over their child's genital surgery, who uprooted themselves and introduced their "new" baby, all the while trying to remember to use the new pronoun when discussing their child's past, would then also have the adaptability to discuss their child's condition and medical history openly and honestly when the time was right to inform the child. And indeed, intersex people who were raised under such conditions have since told life stories that emphasize the insecurity and shame surrounding their parents' decisions, based on doctors' advice, to maintain an important family secret.[104] The time to tell the child never seemed right for many parents, and so children grew up knowing that something was amiss (all the trips to the doctor and the hospital, not to mention the scars, told them that much) but the details remained unexplained.

The sociologist Sharon E. Preves has done research with contemporary intersex adults, confirming that, although attempts at secrecy were perhaps well-meaning, the results were devastating. The people she interviewed uniformly reported that efforts to remain silent about their conditions and their surgeries "only served to enforce feelings of isolation, stigma, and shame—the very feelings that such procedures are attempting to alleviate." Those who needed recurring medical intervention felt this most intensely and used the words "monstrous, Other, and freakish," to describe how they felt about themselves.[105]

By 1968, Money had stopped recommending that families move out of town to avoid disclosure. It was better for parents not to try concealment: "It is emotionally too exhausting to have to live in perpetual apprehension that the skeleton in the family closet, namely the sexual ambiguity, one day will out."[106] In 1981 he wrote, "It is to all intents and purposes impossible for parents to escape the history of a reannouncement of their baby's sex by severing all ties with people and places of the past and reestablishing life anew, incognito." In a study of fourteen babies born with micropenises who were surgically reassigned as girls (at least one of these babies had XY chromosomes but was reassigned anyway), he found that none of the parents relocated; they all simply told people what had happened. Money revised the policy at Johns Hopkins to offer counseling to parents to help them both assimilate the information about their child's condition (so that they could provide informed consent for surgery) and offer the information to family and friends.[107] But evidence from adult intersex patients suggests that doctors did not follow Money's revised suggestions on disclosure or that parents were unable to heed the advice. Medical students in the 1990s continued to learn that it was ethically responsible to withhold information or tell partial truths to spare patients the reality of their diagnosis.[108]

Though the 1950s protocols had advocated a family move following gender-changing genital surgery, they did not recommend universal secrecy. In fact, Money and his colleagues were cognizant of children's intuitive nature and suggested that they would figure things out even if parents tried to conceal the child's medical condition. They urged doctors to tell the truth to affected children. "Thus, far from burdening them with unnecessary worries, it is actually a lifting of the burdens of secret worries and doubts for the doctor to talk frankly with children. Truth is seldom as distressing as the mystery of the unknown," they counseled. Using the metaphor of "unfinished genitals," Money and his team thought it best for a child to learn about his or her body's limitations, such as sterility, than to be confronted with such knowledge later in life when the shock and disappointment might be greater.[109]

The optimal level of disclosure depended on the particular medical condition. Telling girls that they would not be able to have children when they got older did not seem damaging. Money and the Hampsons believed that sterility ranked low on a list of what disturbed female patients with gonadal agenesis (no gonadal tissue), for example. Short stature and lack of menses troubled them far more. Being like the other girls, particularly for teenagers, was most important, they argued; other girls they knew menstruated, but none of them had babies yet. A girl's femininity could still be intact if she could envision herself a mother, the team believed, even if she had to mother an adopted baby rather than one she had conceived. Parents should not hesitate to tell their daughters about their infertility, for a teenage girl could assimilate this well and "accept with equanimity her destiny of adoptive motherhood rather than of biological maternity."[110] Though her "fantasies of childbearing" would have to be modified somewhat, a young girl could still dream of motherhood, which Money believed was an essential component of successful adherence to femininity.

Girls without gonadal tissue should be told, then, that they would not be able to have children, according to Money, Hampson, and Hampson, but if they had male chromosomes as well, then that fact should be concealed from them. If the girls found out this information, they would be hopelessly confused "by present day misconceptions that the entirety of one's sexuality is genetically determined."[111] If there was one thing the team was sure of, it was that a person's sense of herself or himself as female or male was independent of genes and chromosomes. Alleviating confusion and doubt was the only way to make sure that one's gender role remained stable, and sometimes that required secrecy.

John Money's Critics

The case that overtly exposed fallacies in Money's and the Hampsons' reasoning involved a boy, an identical twin, David Reimer, who was not intersexed; his penis had been severely damaged during circumcision when he was seven months old, and his subsequent gender reassignment had been hidden from him. This case, known pseudonymously in the medical literature as the John/Joan case, has been well documented elsewhere, and so the details will not be specified here, except to say that

Money advised the parents to remove the child's injured male organs altogether and raise him as a girl. As the baby was only seventeen months old when the parents adopted this strategy, Money believed that the window of opportunity for sex reassignment was still open and that this child would thrive as a girl, if the parents could successfully raise "her" in an unerringly feminine manner, pink frilly dresses and all. Throughout "her" childhood years, "Brenda" as she was then called by her family, visited Money for psychiatric evaluations at Johns Hopkins, and Money consistently reported that she had effectively adapted to her new life as a girl.[112]

The case became widely disseminated in 1972, when Money and Anke A. Ehrhardt published *Man and Woman, Boy and Girl: The Differentiation and Dimorphism of Gender Identity from Conception to Maturity*. The book, aimed at a lay as well as a medical audience, lauded the surgery on David as well as Money's gender theory more generally. In January 1973, *Time* magazine picked up the story following Money's presentation at the Washington, D.C., meeting of the American Association for the Advancement of Science, and John Money became known as a supporter of women's liberation for proposing that nurture was more important than nature, as the case seemed to demonstrate. His hypothesis that gender roles were learned as opposed to inherent echoed what Simone de Beauvoir had eloquently declared in her 1949 epic feminist manifesto, *The Second Sex*: "One is not born, but rather becomes, a woman."[113] Even though "Brenda" had been born with male genitalia and was not intersexed, the story of "her" supposedly triumphant gender switch catapulted Money's theories into the mainstream. Money wrote hundreds of articles and several books expatiating on his protocols, and most intersexed infants encountered his hypotheses in the practices followed by their own pediatricians and surgeons, in whatever hospital they were born.[114]

"Brenda" was never happy as a girl, though Money had glossed over as passing phases her bouts of aggression and depression during her youth. At fourteen, her therapist encouraged her parents to tell the truth about what had happened. At that point, Brenda decided to live as a boy again and chose the name David. Years later he recalled being told the truth: " Suddenly it all made sense why I felt the way I did. I wasn't some sort of weirdo. I wasn't crazy."[115] David began to try to undo what years

of surgeries and hormones had done to his body; he had a double mastec-
tomy (estrogen had given him breasts) and penile and testicular implants.
Testosterone injections gave him male musculature where there was none
when he was Brenda. At age thirty, he met Milton Diamond, a professor
of anatomy and psychology who had long disagreed with Money's gender
theories. David agreed to participate in a scathing critique of Money's
work published in the *Archives of Pediatrics and Adolescent Medicine* in
1997. Diamond and his coauthor Dr. H. Keith Sigmundson concluded,
"There is no support for the postulates that individuals are psychosexu-
ally neutral at birth or that healthy psychosexual development is dependent upon the appearance of the genitals."[116] Money's hypotheses were
publicly under attack.

Money had his critics even in the earlier years. In response to the *Time*
article that acclaimed the dramatic John/Joan case, one letter writer per-
ceptively asked, "Isn't it typical of this society that doctors should convert
an infant with deformed genitalia into a female 'with the realization that
he could never be a normal man.' The fact that 'she' could never be a
normal woman either does not seem to bother the sex experts."[117] And as
early as 1965, Diamond had critiqued Money's reasoning, questioning the
theory's uncanny rise to prominence without serious challenge. Diamond
contended that Money's data could bear a different interpretation and
that there was no evidence to corroborate his notion that sex roles were
undifferentiated and neutral at birth. Perhaps one could conclude from
the data that intersex people "are flexible when it comes to the assumption of an incongruous sex role." But to assume that a sex role is without
"biological prepotency or prenatal organization and potentiation is unjus-
tified and, from the present data, unsubstantiated."[118]

In 1970 Dr. Bernard Zuger, a New York psychiatrist, also contested
Money's theories. He reviewed the methodology of Money and the Hampsons, reinterpreted their data, and collected material from other cases that
suggested that many intersex people did experience a "change of sex without hazardous psychological consequences" later in life and despite their
rearing.[119] Zuger concluded that sex assignment and rearing did not over-
ride all other determinants of gender for the intersexed or for anyone else.
Money disagreed, and in a rebuttal he disputed Zuger's research methods
and sneered, "It is difficult for the seeing to give art instruction to the
blind."[120]

Intersex activists are currently challenging Money's presumptions and protocols. Beginning in the 1990s, intersex adults and feminists have been working to improve the medical and social attitudes toward people with intersex, focusing in particular on ending the optimum gender of rearing model.[121] They argue that the so-called normalizing genital surgery that most infants born with unusual genitals endure does not cure the intersex condition and is often simply cosmetic. Sometimes, as intersex scholar Morgan Holmes has pointed out, surgery to alleviate genital ambiguity might actually have the opposite effect: when genetic boys with micropenises are fashioned into girls they are "actually *made into* intersex persons." "In fact," she argued, "it seems that such a child, in a female assignment is actually more intersexed following surgery than at the point of diagnosis."[122] Allegedly corrective surgery sometimes does more harm than good by causing incontinence and permanent loss of sexual sensation. As Alice Dreger, an academic historian and past chair of the board of directors of the former Intersex Society of North America (ISNA), put it, "Why perform irreversible surgeries that risk sensation, fertility, continence, comfort, and life without a medical reason?" ISNA's (and now Accord Alliance's) efforts are making headway, but they have not yet been universally accepted. At a 2004 meeting of the Section on Urology of the American Academy of Pediatrics, several leaders in the field cautioned against invasive cosmetic surgeries, while others continued to advocate early aggressive operations.[123]

At the very least, parents to whom an intersexed child is born today can easily find ISNA's Web site and be introduced to alternatives that parents of children born in the 1950s, 1960s, 1970s, and 1980s hardly encountered. Even a second medical opinion at that time would probably have elicited the same advice, based on the popularity and reputation of Money's work. According to Anne Fausto-Sterling, Money's theory of psychosexual development was considered "the most progressive, most liberal, most up-to-date point of view around."[124] Some of his high repute was of Money's own making. In his 1981 article about boys having their micropenises removed and being successfully reassigned as girls, he referred to the 1973 *Time* magazine article that originally announced this procedure in the John/Joan case. He mentioned a devoutly religious mother who agonized over her seventeen-month-old infant, born with a micropenis. When a neighbor showed her the article about Money, "This article represented divine intervention," he boasted. It led her to have her baby reevaluated at Johns Hopkins and to opt for a sex reassignment, even though the baby was nearing the end of the crucial window for a gender switch, according to Money's own arguments. Money credited the parents' "superior ability to resolve the crisis of their child's sex, as soon as they received the help they needed" and claimed that "the 10-year outcome of the case is successful."[125]

If Francis Wharton and Moreton Stillé, authors of a nineteenth-century medical jurisprudence text, had been alive, they would no doubt have disapproved of Money's approach to surgical intervention on infants' genitals. Recall from chapter 2 that they had vehemently opposed the first such "corrective" surgery, by Dr. Samuel D. Gross, who had removed the testicles of a three-year-old girl patient on the theory that removal would prevent the pangs of an unfulfilled libido. Wharton and Stillé had insisted that genital surgery "removes merely the external."[126] Their prophetic warning that "normalizing" surgery (other than lifesaving procedures such as surgery to ensure adequate voiding) does not eliminate intersex has, more often than not, gone unheeded in the intervening hundred-plus years.

Today's intersex activists have a voice that is largely absent from the earlier historical record. Like others influenced by the 1970s health movement, they seek appropriate patient-centered medical care, support groups for parents, and transparency and disclosure by physicians. They challenge the insensitivity that attends medical photography and medical observation, and they demand that broad-minded, evidence-based decisions govern their medical care. What parents and patients want is not inconsistent with contemporary medical practice; faced with virtually any other medical condition, less fraught with controversies over sex and gender, doctors strive to meet such expectations. Yet, as I hope this book has shown, the medical world's hearkening to that voice will represent a radical departure from the past.

Divergence or Disorder?

The Politics of Naming Intersex

Iₙ THE TIME THAT IT TOOK to write this book, the term "intersex" has come under scrutiny and is the subject of much debate from many quarters. The epilogue will analyze the current controversy, placing it in the historical context of over three hundred years of intersex management in this country. Though the disagreement centers on what to call "intersex," its ramifications are much more than lexicological. Indeed, the debate underscores the central dilemma of this book: the evolving perception of atypical bodies, particularly bodies that raise our anxiety level because they seem to muddle clear gender divisions. Readers will come to see, I hope, how our contemporary standpoint is no less fraught with cultural biases than our predecessors'. Understanding the naming issue from an informed historical perspective may help us to see how we have arrived at this point and where we might go from here.

As a historian, I am accustomed to thinking about change over time, and I know that change often happens slowly. Not so with the recent nomenclature shift in the world of intersex. In medical settings, many of the conditions previously grouped under the broad categories of "intersex" and "hermaphroditism" are now generally being called "disorders of sex development" (DSDs). The new term was agreed upon in October 2005 at a conference hosted by the Lawson Wilkins Pediatric Endocrine Society and the European Society for Paediatric Endocrinology (hereafter called the Chicago Consensus Conference) and is quickly becoming ubiquitous. Though participants at the Chicago conference reached the decision to change the nomenclature by consensus, it has not been universally embraced. Each of these three terms—"intersex," "hermaphrodite," and "disorders of sex development"—is controversial and divisive. Here I suggest a modified new term, "divergence of sex development." Having surveyed the history of medical management of intersex conditions and recognizing the distrust it has created, I hope that you agree that using "divergence of sex development"—and eschewing the loaded word "disorder"—might reduce conflict and satisfy intersex people, their parents, and physicians.

How to name diverse conditions involving aspects of external genitalia, sex chromosomes, internal reproductive anatomy, and gender identity raises political as well as medical questions. The choice of nomenclature influences not only how doctors interpret medical situations but, also and as important, how parents view their affected children, how intersex people understand themselves, and how others outside medical settings—such as gender and legal scholars, historians, and media commentators—think, talk, and write about gender, sex, and the body.

The three terms, "hermaphrodite," "intersex," and "disorders of sex development," might seem synonymous, but there are significant differences, and their use has controversial consequences. "Hermaphrodite" and "hermaphroditism," as we have seen, are archaisms that can still be found in medical writings, but they are vague, demeaning, and sensationalistic. Historically, "hermaphrodite" has been one of the more neutral descriptors; derogatory terms such as "freak of nature," "hybrid," "impostor," "sexual pervert," and "unfortunate creature" pervade early medical literature. In an 1842 article on malformations of the male sexual organs, for example, one doctor referred to "these mortifying and disgusting imperfections."[1] As the word "hermaphrodite" continues to evoke images of mythical creatures, perhaps even monsters and freaks, it is not surprising that people would want to avoid the label.

Starting in the early 1990s, activists instead advocated "intersex" (first introduced by the biologist Richard Goldschmidt in 1917) to describe the set of conditions previously called hermaphroditism—namely, discordance between the multiple components of sex anatomy.[2] The Intersex Society of North America (ISNA), founded by the intersex activist Cheryl Chase in 1993, sought to erase the stigma perpetuated by negative labeling and support those with congenital conditions that fall under that rubric. While most often using "intersex" to refer to themselves, some of the affected as well as their supporters also consciously reclaimed the term "hermaphrodite," co-opting the negative label in a bold effort to call attention to their concerns and to dispel pathological connotations associated with these conditions.[3] "Intersex" thus took on a political valence, as many of the intersexed proudly sported t-shirts that proclaimed themselves "Hermaphrodites with Attitude" and, wearing these shirts, protested at medical conferences against stigmatization and unnecessary infant genital surgeries.[4]

Some parents, though, were uncomfortable with the "intersex" label for their affected children. To them, "intersex" meant a third gender, something in between male and female. They wanted to see their newborn babies as girls or boys, not as "intersex." Even though intersex activists advocate raising children as girls or boys rather than as a third, in-between category, some parents found the label as frightening, off-putting, and freakish as "hermaphrodite." Dr. Arlene Baratz, for example, has affirmed that as a parent of a daughter with androgen insensitivity syndrome, she was "shocked and unnerved" when she first confronted the words "intersex" and "pseudohermaphrodite." She has said that other parents in her support group have also rejected the label "intersex" because of its implications.[5] Others have associated the word "intersex" with sexuality, eroticism, or sexual orientation and have preferred to discuss their child's anatomical condition without focusing on his/her future sexual activities.[6]

For their part, doctors have never fully incorporated "intersex" into their vocabulary. There has never been agreement on what "intersex" means or on what conditions constitute intersex. Lacking a suitable alternative, many physicians still use (or until very recently have used) the nineteenth-century terms based on "hermaphrodite," including "male pseudohermaphrodite" (having ambiguous external genitalia with male gonads), "female pseudohermaphrodite" (having ambiguous external genitalia with female gonads), and "true hermaphrodite" (having both testicular and ovarian tissue regardless of external characteristics). The continued use of the term "hermaphrodite" has humiliated some patients and embarrassed some parents. Some physicians have used the hermaphrodite label in their medical records but avoided saying the word in front of children or parents, for fear it would cause alarm and harm. That awkward circumspection, while well meant, has contributed to a penumbra of secrecy and shame surrounding these conditions.[7]

Supporters of the relatively new alternative term "disorders of sex development" believe it deemphasizes the identity politics and sexual connotations associated with "intersex" and the degradation associated with "hermaphrodite" and instead draws attention to the underlying genetic or endocrine factors that cause prenatal sex development to take an unusual path. Many proponents of the name change believe that using "DSD" could improve medical care for affected children and health-care workers' interactions with their families because it avoids sensationalizing health conditions, allowing doctors to focus solely on therapeutic issues.[8] Of course, each specific condition has its own terminology and its own protocol, but thinking across diagnoses has advantages, such as encouraging clinicians to collaborate in multidisciplinary teams when they work with patients faced with similar medical and psychosocial needs.

The 2005 Chicago Consensus Conference, composed of fifty invited international experts in the field (principally M.D.s), included only two intersex adults and no parents of affected children.[9] Yet, despite that limitation, the conference was pathbreaking in its inclusion of any intersex adults in the policy-making process. Participants came to other important agreements besides the name change, including the need for more open communication between doctors, patients, and families and for a more conservative approach to surgery. Perhaps most importantly, the consensus statement acknowledged that there is little evidence that infant genital surgery does what it has been assumed to do: improve attachment between child and parents, ease parental distress about atypical genitals, ensure gender-identity development in accordance with the assigned gender, or eliminate the intersex condition.

The term "disorders of sex development" may promote clarity for doctors who diagnose patients with such conditions and provide some relief for patients and parents, but it has produced rancor among some adults who identify as intersex. Specifically, they reject the word "disorder." The disability-rights movement has taught us that atypicality does not necessarily mean disorder. Doesn't "disorder" imply that something is seriously wrong and needs to be corrected?

If the word "disorder" connotes a need for repair, then this new nomenclature contradicts one of intersex activism's central tenets: that unusual sex anatomy does *not* inevitably require surgical or hormonal correction. ISNA (the former primary activist organization) advocated eliminating infant or childhood genital surgical procedures other than those that are lifesaving.[10] Using the word "disorder" elides a crucial point that some of these surgeries, such as clitoral recession, serve primarily social rather than medical goals. As the scholar Suzanne J. Kessler declared, "Gender ambiguity is 'corrected,' not because it is threatening to the infant's life, but because it is threatening to the infant's culture."[11]

More broadly, should we think of intersex bodies as disordered when they actually are more common than most people know? Since at least one out of every two thousand babies is born with such anatomy, perhaps it makes more sense to think of this phenomenon as part of the natural, albeit unusual, spectrum of human conformation.[12] The label "disordered" marks an individual as patently impaired, with a body that needs to be poked and prodded until it fits neatly into the recognizable binary categories of female and male.[13] Using the word "disorder" thus contradicts the central precept of disability politics, which asserts that difference need not be seen as inherently insufficient or defective.[14]

The idea that culture, rather than the body, needs to be changed has informed feminist scholars as well as disability theorists.[15] The notion that biological sex may not be as rigidly binary as conventionally thought has appealed particularly to academic feminists.[16] In turn, some intersex people have benefited from the feminist understanding of the complicated relationship between gender and sex. Some have corroborated the feminist supposition that we should think of sex, like gender, as on a continuum, as more flexible than strictly female or male. Some have felt more comfortable identifying not solely either as female or as male, but as intersexed, with a combination of physical and behavioral characteristics.[17] From the perspective of sexual politics, then, shifting from "intersex" to "disorders of sex development" represents a denial of a core feminist and intersex-activist principle that sex and gender are fluid.

Suspicion among intersex activists about the medical management of intersex is based on their knowledge of the long, painful history of doctors' approach to intersex people. Ever since the early nineteenth century, when doctors began to professionalize and publish their cases in medical journals, we can trace not only their cruelly judgmental descriptors of these conditions and people but also the damaging therapeutic treatment they have dispensed. The ways intersex bodies have been scrutinized and pathologized have been negative, harmful, and based not on medical necessity but on social anxieties about marriage and heterosexuality and on the insistence on normative bodies.[18] The prevention of homosexuality has long motivated surgical and nonsurgical sex assignment in this country, for example, and even today the use of the prenatal drug dexametha-sone as a treatment for congenital adrenal hyperplasia (a condition that affects the adrenal glands and can cause masculinization in girls), may be linked to its deterrence.[19] Those with atypical genital anatomies have had their bodies reshaped and sculpted to look (and presumably to act) more typical, even though evidence suggests that those who underwent such life-altering surgery have not had more successful outcomes and happier lives than those who avoided it.[20]

This long history of medical denigration and experimentation is one reason some intersex activists are wary of having a new, medicalizing term thrust upon them. At a recent ISNA-sponsored symposium, during the annual Gay and Lesbian Medical Association meeting in October 2006, Peter Trinkle, the board president of Bodies Like Ours, an intersex edu-cational and peer-support organization, voiced the complaints of a wide spectrum of intersex activists who contend that the new term further stigmatizes and pathologizes their lives. Using "disorders of sex devel-opment," Trinkle argued, suggests that the ends of sexual development should be "normal" male or female bodies, whereas the term "intersex" seems to imply and accept biological diversity. Others worried that the use of pathologizing medical terminology will overshadow the political progress and advances in gender and disability theory made since the early 1990s.[21]

Where does this leave us? Some intersex people (though not all) do not want their conditions to be pathologized as "disorders." Parents of af-fected children do not want them to be considered "abnormal," or inter-sexed. They want "normal" girls and boys. For their part, doctors want to provide the best care possible, and, ironically, in their world labeling something a "disorder" normalizes it. Doctors (and insurance companies) understand disorders.

Hence the dilemma: Whose naming should/will prevail? Another way of posing this question is: Whose needs should be met? Most speakers at ISNA's October 2006 symposium—including doctors, social workers, therapists, intersex adults, and parents—agreed that using the term "DSD," despite its limitations, would likely benefit infants and children. Indeed, that was why in 2005 an independent group of intersex people, parents, and clinicians (known as the Consortium on the Management of Disorders of Sex Development) organized by Alice Dreger as ISNA's director of medical education used the term "DSD" in their new clinical guide-lines and parents' handbook—because they believed it would advance "patient-centered care."[22] If it takes using this new term to get medical professionals to listen to what those affected by medical decisions want, then perhaps the end justifies the means. Most of the authors of the consortium's two books hold this pragmatic position, though even in that group, three of the contributors objected to the use of the term and added disclaimers to the consortium Web site.[23]

By adopting the term "disorders of sex development" and granting doctors the power to do the naming, do we, in fact, give disproportionate control to the medical establishment? I think so, even though using "DSD" in the medical arena does not preclude intersex people from claiming the term "intersex" for themselves, as adults. One speaker at the 2006 ISNA symposium who has androgen insensitivity syndrome explained that she used to recoil from the intersex label because she thought it was just a euphemism for "hermaphrodite." Now that DSD is available, she is better able to incorporate intersex into her identity; consequently, she thinks of herself as a woman who is intersexed because she has a DSD.

My suggestion is to retain the acronym DSD, but have it stand for "divergence of sex development." Divergence of sex development would be less pathologizing than disorders of sex development and yet would satisfy those who want to minimize the emphasis on genitals, gender identity, and sexual orientation that the "intersex" label may encourage. The use of "divergence" would not label intersex people as in a disordered physical state in unquestionable need of repair. "Variations of sex development" has been suggested as a nonstigmatizing label, but the term raises objections because VSD is already a medical acronym (a Google search for VSD brings up pages on ventricular septal defect) and also because a "variation" might downplay the seriousness of some intersex conditions such as congenital adrenal hyperplasia.[24]

The divergence of sex development designation would neither prohibit medical intervention nor inevitably demand it. It is true that "divergence" still implies a departure from a typical developmental path, but in terms of incidence, the word is consistent with reality. As with any divergence, doctors would be compelled to see if there was a serious underlying organic health problem requiring intervention and then, using the same principles of medical ethics applied to other conditions, make therapeutic choices accordingly, without imparting a sense of shame and stigma to the parents or the patients.

While the term "intersex" can still be a useful political expression for adults, using "divergence of sex development" in the medical sphere would allow physicians to evaluate intersex issues in all their complexity, and that is in everyone's best interest. I am not so naïve as to think that change among doctors or insurance companies (which might pay only for treatments of "disorders") will come easily. But at the very minimum, the new term could sensitize clinicians to the implications of the language they use. With nonstigmatizing, non-correction-demanding nomenclature, doctors, therapists, and parents can assess those affected more responsibly and ethically—the desired goal in this, and indeed in any, medical situation.

Recently, a woman gave birth to a baby whose sex could not be immediately determined. The infant's parents were friends of a friend, and so, a few hours after delivery, I found myself talking on the phone with a perplexed and concerned new father searching for information and guidance. What could I tell him as a historian and a friend? During the pregnancy the prospective parents had been told that their baby would be a girl. Pink paraphernalia filled their house. He wondered, what should they now tell family and friends when asked if their newborn was a boy or a girl? Every child is unique, sometimes in ways we can hardly imagine. Yet, I told him, he and his wife were not the first parents to have an intersex baby, meaning a child with discordance among some of the multiple components of sex anatomy: internal reproductive organs, external genitalia, chromosomes, and hormones. No matter what their baby's genitals looked like, or what gender the baby turned out to be, they would not be alone; there would be other parents to talk to. I offered him my copy of Accord Alliance's *Handbook for Parents*, originally published by ISNA.[25] After learning that the doctors said the baby was otherwise healthy and not facing any immediate medical emergencies, I encouraged the father to wait and see what the doctors deduced from chromosomal and hormonal tests but not to agree to any genital surgery, should it be suggested. Don't do anything that cannot be undone, I recommended.

Knowing they would be in the hospital for a few days with nothing definitive to report, the parents bravely and sensibly decided to tell the truth—that they did not yet know if their baby would be a boy or a girl. That simple and straightforward response represents a revolutionary shift in the history of intersex births. For hundreds of years these ambiguous bodies had been considered a punishment of an angry God, the fault of a mother's deranged imagination, or—at best—a shameful disfiguration. For the last half-century, the medical establishment, if more enlightened, has nevertheless typically advised parents to keep their child's condition secret, even to relocate and sever social and familial ties in the event that their child's gender was changed surgically during infancy. Often, the facts of birth conformation and alteration were to be concealed even from the affected children themselves. Intersex activists have campaigned to end secrecy and the shame consequent upon concealment, and their message seems to be taking hold. My friends told the truth about their baby's condition and received nothing but support from friends and family.

The doctors remained uncertain after a few days of testing. The baby's chromosomes were mixed; they included 68 percent XY (typical of males), 30 percent XO (a missing second sex chromosome), and 2 percent XYY (an extra male chromosome, a phenomenon that is estimated to occur in one in a thousand births). Hormone tests brought similarly ambiguous results, though the doctors were fairly confident that the baby's elevated levels of testosterone and low levels of estrogen meant that he should be raised as a boy. The parents decided to give the baby a gender-neutral name, in case the doctors were wrong and their child felt more like a girl than a boy as he grew older.

Neither the family's pediatrician nor the pediatric urologist suggested immediate infant genital surgery. A few months later the baby did undergo exploratory surgery to see if a mass in his abdomen was an undescended testicle, or an ovary, or an ovotestis (a gonad containing both ovarian and testicular tissue). Surgeons found in his abdomen a half-formed uterus on his right side, which they removed, and a testicle capable of producing testosterone on the left. The baby will also have surgery to detach the attached shaft of his penis from his body and then another surgery for hypospadias repair, to reroute the urethra through his penis.

A generation ago this baby might have been assigned female, despite his chromosomes and hormone levels, because it seemed easier to create female anatomy out of ambiguous genitals than to raise a boy with an imperfect penis.[26] Despite all the barriers to changes in medical protocol, which Katrina Karkazis astutely details in her recent book, *Fixing Sex: Intersex, Medical Authority, and Lived Experience*, activists have made great strides in persuading doctors to think twice about what they advocate for intersex patients; some physicians have begun to confront their bias toward creating normative bodies.[27]

On the other hand, infant surgery to normalize genital appearance is still being promoted, even without conclusive evidence that the child will be better adjusted or happy about it as he or she grows older. "Conclusive evidence" that would justify surgery hardly exists in medicine. Which surgeries are judged successful and by whose standards? Surgeons might be satisfied by the appearance of their patient's genitals while the patient might express dissatisfaction. Surgical techniques are always improving, but there are still risks, including the formation of scar tissue, risk of incontinence, diminished sexual sensation, and the need for further surgeries. Some intersex activists would like to bar surgery until after puberty, when sexual development might clarify the situation and when the person affected is of an age to choose what, if anything, should be done. Aware of the tragedies of the past, which as we have seen were often exacerbated by assumptions about "normal" male and female genitals, bodies, and sexual behaviors, activists wish to at least accord the possible surgical candidate a voice.

The birth of my friends' baby has demonstrated to me the ambiguous nature of intersex medical treatment at this particular moment. When set in the context of a broader American history it is easier to see how decisions made today, despite progress, remain difficult and contingent. Today's clinicians and parents want to do right by their patients and children, but that goal does not make their deliberations any easier. The optimum protocol for bodies in doubt therefore is for physicians to be mindful that their medical views are embedded in a particular context. Doctors should do today what many avoided historically: be explicit about their biases and admit uncertainty. Such honesty might alleviate some of the damage born of hubris such as we have seen in the past and encourage more cautious and judicious care.

NOTES

Preface

1. Alice Domurat Dreger's pioneering book, *Hermaphrodites and the Medical Invention of Sex,* compares the history of British and French intersex cases. There are recent works in sociology, anthropology, and biology, such as Suzanne J. Kessler, *Lessons from the Intersexed* (New Brunswick, N.J.: Rutgers University Press, 1998); Anne Fausto-Sterling, *Sexing the Body: Gender Politics and the Construction of Sexuality* (New York: Basic Books, 2000); Katrina Karkazis, *Fixing Sex: Intersex, Medical Authority, and Lived Experience* (Durham, N.C.: Duke University Press, 2008); and Sharon E. Preves, *Intersex and Identity: The Contested Self* (New Brunswick, N.J.: Rutgers University Press, 2003); these studies, however, focus on the twentieth century, commencing where my story ends.

2. The word "hermaphrodite" conjures up images of mythical creatures, perhaps even monsters, and so it is not surprising that intersex activists began to call for its elimination in the early 1990s. They instead advocated the term "intersex" to describe the set of conditions formerly known as hermaphroditic. Today many in the medical community are using the acronym DSD (disorders of sex development) instead of "intersex," a controversial change of nomenclature, which I discuss (and suggest an alternative for) in the book's epilogue.

3. For three recent examples of books that examine the confluence of medicine and culture, see Lennard J. Davis, *Obsession: A History* (Chicago: University of Chicago Press, 2008); Barron Lerner, *The Breast Cancer Wars: Hope, Fear, and the Pursuit of a Cure in Twentieth-Century America* (New York: Oxford University Press, 2003); and Judith Houck, *Hot and Bothered: Women, Medicine, and Menopause in Modern America* (Cambridge: Harvard University Press, 2006).

4. Hypospadias is a condition whereby the urinary opening of the penis falls somewhere other than at the tip. The severity of hypospadias can vary; sometimes it is such that the penis resembles labia. Turner syndrome is a chromosomal condition in which girls have only one X chromosome. Congenital adrenal hyperplasia involves the inability of the adrenal glands to produce cortisol, resulting in an overproduction of other hormones and masculinization in girls. Androgen insensitivity syndrome in a genetic condition in which the body is unable to respond to androgens. AIS women develop as girls but have XY chromosomes and unde-

scended testes and a short vagina or no vagina. 5-alpha-reductase deficiency is an enzyme deficiency that causes people who appear to be girls at birth to virilize at puberty. The protagonist in Jeffrey Eugenides' novel, *Middlesex,* had this condition. Sex chromosome mosaicism is characterized by a mixture of sex chromosomes, which can result in ambiguous reproductive organ development. These are just a few examples of conditions that could be classified as intersex.

5. The figures on the incidence of intersex range widely, depending on which conditions one includes in the calculation. One incident in 2,000 births represents a moderately inclusive estimate. Estimators who are more selective in determining what characteristics qualify as intersex put the figure at 2 in 10,000. Researchers at Brown University recently stated that the frequency of people receiving "corrective" genital surgery is between 1 and 2 per 1,000 live births. The Chicago Consensus Conference put the figure at 1 in 4,500. See I. A. Hughes, "Consensus Statement on Management of Intersex Disorders," *Archives of Disease in Childhood* 91 (2005): 554–63; Anne Fausto-Sterling, *Sexing the Body,* 76; Melanie Blackless et al., "How Sexually Dimorphic Are We? Review and Synthesis," *American Journal of Human Biology* 12 (2005): 151–66; and Dreger, *Hermaphrodites,* 40–43.

6. John W. Money, "Hermaphroditism: An Inquiry into the Nature of a Human Paradox" (PhD diss., Harvard University), midyear, 1951–52.

7. Pediatric urologist Justine Schober writes, "As surgeons, we have addressed the aesthetic appearance and functionality of the external genitalia with the belief that these physical changes we impose would help increase psychosocial and psychosexual comfort . . . The immediate aesthetic results seem to continually improve. However, the long-term efficacy of the structural results of various surgeries and their impact on the individuals' psychological, social, and physical adjustment remains unknown." See Justine Murat Schober, "A Surgeon's Response to the Intersex Controversy," *Journal of Clinical Ethics* 9, no. 3 (1998): 393–97, quotation on 393.

Chapter One: Hermaphrodites, Monstrous Births, and Same-Sex Intimacy in Early America

Epigraph. Dr. Alexander Hamilton, *Hamilton's Itinerarium; Being a narrative of a journey from Annapolis, Maryland, through Delaware, Pennsylvania, New York, New Jersey, Connecticut, Rhode Island, Massachusetts and New Hampshire, from May to September, 1744,* ed. Albert Bushnell Hart (St. Louis, Mo.: Printed only for private distribution by W. K. Bixby, 1907). Electronic Edition, Maryland Institute for Technology in the Humanities.

1. On Hamilton's attitude toward women and their "proper" place in the

female sphere, see Elaine G. Breslaw, "Marriage, Money, and Sex: Dr. Hamilton Finds a Wife," *Journal of Social History* 36, no. 3 (Spring 2003): 657–73.

2. According to historian Abraham Luchins, at the time of the American Revolution, only 400 of the estimated 3,500 doctors in America even held medical degrees. See Luchins, "Social Control Doctrines of Mental Illness and the Medical Profession in Nineteenth-Century America," *Journal of the History of the Behavioral Sciences* 29 (January 1993): 39; and Charles E. Rosenberg, "The Therapeutic Revolution: Medicine, Meaning, and Social Change in Nineteenth-Century America," *Perspectives in Biology and Medicine* 20 (1977b): 485–506.

3. On monstrosity in the early modern world, see Katherine Park and Lorraine Daston, *Wonders and the Order of Nature, 1150–1750* (New York: Zone Books, 1998); and Marie-Hélène Huet, *Monstrous Imagination* (Cambridge, Mass.: Harvard University Press, 1993).

4. See David D. Hall, *Worlds of Wonder, Days of Judgment: Popular Religious Belief in Early New England* (New York: Alfred A. Knopf, 1989).

5. Richard S. Dunn, James Savage, and Laetitia Yeandle, eds., *The Journal of John Winthrop, 1630–1649* (Cambridge: Harvard University Press, 1996), 254.

6. *Journal of John Winthrop*, 254–55. On the signs that indicated the devil's presence, see Carol F. Karlsen, *The Devil in the Shape of a Woman: Witchcraft in Colonial New England* (New York: W. W. Norton, 1987), 16–17.

7. Pietro Martire d'Anghiera, *The Decades of the Newe Worlde or West India*, ed. and trans. Richard Eden (London, 1555), in *The First Three English Books on America, 1511–1555 A.D.*, ed. Edward Arber (Birmingham, 1885; New York: Kraus, 1971), 53. For a discussion of Winthrop's interpretation of monstrous births as heavenly portents, see Robert Blair St. George, *Conversing by Signs: Poetics of Implication in Colonial New England Culture* (Chapel Hill: University of North Carolina Press, 1998), 169–73. See also K. Park and L. J. Daston, "Unnatural Conceptions: The Study of Monsters in Sixteenth-and Seventeenth-Century France and England," *Past and Present* 92 (1981): 20–54; and Johan Winsser, "Mary Dyer and the 'Monster' Story," *Quaker History* 79, no. 1 (Spring 1990): 20–34.

8. David D. Hall, ed., *The Antinomian Controversy, 1636–1638: A Documentary History* (Durham, N.C.: Duke University Press, 1990), 214.

9. Aristotle [pseudonym], *Aristotle's Master-Piece; or, The Secrets of Generation Displayed in All the Parts Thereof . . . Very Necessary for All Midwives, Nurses, and Young-Married Women* (London: printed for W. B., 1694); see Mary E. Fissell, "Hairy Women and Naked Truths: Gender and the Politics of Knowledge in *Aristotle's Masterpiece*," *William and Mary Quarterly*, 3rd ser., 60, no. 1 (January 2003): 43–74. Fissell notes that the book went into more editions than all other books on the subject combined. On the differences between editions

and the book's popularity, see Vern L. Bullough, "An Early American Sex Manual, or Aristotle Who?" *Early American Literature* 7, no. 3 (Winter 1973): 236–46; and Helen Lefkowitz Horowitz, *Rereading Sex: Battles over Sexual Knowledge and Suppression in Nineteenth-Century America* (New York: Alfred A. Knopf, 2002).

10. *Aristotle's Master-Piece,* 1694 edition, 180–81.

11. Ibid., 17.

12. By the early nineteenth century, doubt about the significance of maternal imagination on unusual births began to creep into doctors' accounts. While some held on to this explanation, others denied it, though the tone of many articles suggests that the idea had not been completely dispelled, even among medical men. See, for example, Thomas Close, "Singular Monstrosity," *Boston Medical Intelligencer* 3, no. 18 (September 13, 1825): 71. For stories that corroborate the theory, see "Extraordinary Phenomenon," *Pittsburgh Recorder* 1, no. 23 (June 27, 1822): 367; and G. W. Garland, "Acephalous Monster," *Western Journal of Medicine and Surgery* 7, no. 3 (March 1851): 212–14. The idea flourished again in the late nineteenth century, albeit in the context of denying its reality. See, for example, Michel Middleton, "Cases of Malformation: With Reflections on Congenital Abnormalities," *American Journal of the Medical Sciences* 55, no. 109 (January 1868): 69–76. Case VII insists on the coincidental nature of abnormal births, using an example that most readers might interpret as a perfect case of maternal imagination affecting the birth of a child. The incident involved a married black woman who had copulated with a white man, unbeknownst to her husband. Her husband had six fingers on each hand, and by sheer will of her anxiety about being found out, the article suggests, so too did her child. Here the racial "othering" of monstrosity and the theory of maternal imagination commingled.

13. Jane Sharp, *The Midwives Book; or, the Whole Art of Midwifery Discovered,* ed. Elaine Hobby (1671; New York: Oxford University Press, 1999), 92.

14. The direct antecedent to the *Master-Piece* image is a woodcut in the English translation of Ambrose Paré, *The Workes of That Famous Chirurgion Ambrose Parey Translated out of Latine and Compared with the French,* by Thomas Johnson (London: printed by Th: Cotes and R. Young, 1634). See Fissell, "Hairy Women and Naked Truths," 13–14.

15. *Aristotle's Master-Piece, Completed in Two Parts: The First Containing the Secrets of Generation in All the Parts Thereof* (London, 1700), 38. The caption under the picture of the hairy maid says, "The Effigies of a Maid all Hairy, and an Infant that was Born Black, by the Imagination of their Parents" (39).

16. Ibid., 40. On the relationship between menstruation and intercourse, see

Patricia Crawford, "Attitudes to Menstruation in Seventeenth-Century England," *Past and Present* 91 (May 1981): 47–73.

17. *Aristotle's Master-Piece,* 1700 edition, 35. Crawford also discusses the prohibition against intercourse during menstruation. See Crawford, "Attitudes to Menstruation," esp. 63–65.

18. Samuel Farr, "Elements of Medical Jurisprudence" (London, 1788), reprinted in Thomas Cooper, ed., *Tracts of Medical Jurisprudence* (Philadelphia, 1819), 1–80. Farr's book was also excerpted in *Eclectic Repertory and Analytical Review, Medical and Philosophical* 5, no. 4 (October 1815): 11.

19. *Aristotle's Master-Piece,* 1700 edition, 41.

20. James Parsons, *A Mechanical and Critical Enquiry into the Nature of Hermaphrodites* (London: J. Walthoe, 1741), xvi, lii.

21. Nathaniel Shurtleff, ed., *Records of the Colony of New Plymouth in New England. Court Orders,* VI (1678–91) (Boston: William White, 1856), 191–92.

22. John F. Cronin, ed., *Records of the Court of Assistants of the Colony of the Massachusetts Bay, 1630–1692* (Boston: Published by the County of Suffolk, 1928), 131–32.

23. On New England divorce, see Cornelia Dayton, *Women before the Bar: Gender, Law, and Society in Connecticut, 1639–1789* (Chapel Hill: University of North Carolina Press, 1995), esp. 105–56. A law passed in Philadelphia in 1815 stated "that if either party, at the time of the contract, was and still is naturally impotent, or incapable of procreation, it shall and may be lawful for the innocent and injured person to obtain a divorce." See John Purdon, Esq., ed., *A Digest of the Laws of Pennsylvania from the Year One Thousand Seven Hundred to the Twenty-First Day of May, One Thousand Eight Hundred and Sixty-One* (Philadelphia: Kay and Bro., 1862), 345.

24. Theodric Romeyn Beck and John B. Beck, *Elements of Medical Jurisprudence* (Philadelphia, 1838), 104–5.

25. Contemporary medical experts estimate that one out of every two thousand babies is born with an intersex condition, some of which appear at puberty or go undetected until much later, and so the incidence of intersex may even be greater. See the Intersex Society of North America Web site for comprehensive resources: www.isna.org.

26. For two interpretations of Hall's life, see Mary Beth Norton, *Founding Mothers and Fathers: Gendered Power and the Forming of American Society* (New York: Alfred A. Knopf, 1996), 183–97; and Kathleen Brown, " 'Changed . . . into the fashion of man': The Politics of Sexual Difference in a Seventeenth-Century Anglo-American Settlement," *Journal of the History of Sexuality* 6 (1995): 171–93. The case is also discussed in Alden Vaughan, "The Sad Case of Thomas(ine) Hall," *Virginia Magazine of History and Biography* 86 (1978): 146–

48; and Jonathan Ned Katz, *Gay/Lesbian Almanac: A New Documentary* (New York: HarperCollins, 1983), 71–72.

27. H. R. McIlwaine, ed., *Minutes of the Council and General Court of Colonial Virginia, 1622–1632* (1670–76; Richmond, Va.: Colonial Press/Everett Waddey, 1924), 194–95, quotation on 194.

28. See Elizabeth Reis, *Damned Women: Sinners and Witches in Puritan New England* (Ithaca, N.Y.: Cornell University Press, 1997), 112–16.

29. On Jewish law regarding hermaphrodites, see Rabbi Alfred Cohen, "Tumtum and Androgynous," *Journal of Halacha and Contemporary Society* 38 (Fall 1999): 62–85; "Tumtum" refers to those whose sex is indeterminate; "androgynous" refers to those whose organs have both male and female characteristics. See also Sally Gross, "Intersexuality and Scripture," *Theology and Sexuality* 11 (1999): 65–74.

30. Parsons, *Mechanical and Critical Enquiry*, 1–2. On the English classification of hermaphrodites as male or female depending on which sex predominates, see also Catherine Damme, "Infanticide: The Worth of an Infant under Law," *Medical History* 22, no. 1 (January 1978): 1–24. Parsons' advice to choose one sex echoed that of Ambose Paré, the sixteenth-century French surgeon, but with one significant distinction. Unlike Paré, Parsons did not advocate death for a person who switched their gender. Paré had written, "the Laws command those to chuse the sex which they will use, and in which they will remain and live, judging them to death if they be found to have departed from the sex they made choice of, for some are thought to have abused both and promiscuously to have had their pleasure with men and women." See Ambrose Paré, *The Works of That Famous Chirurgeon Ambrose Parey*, trans. Thomas Johnson (London, 1678), 55.

31. Parsons, *Mechanical and Critical Enquiry*, xxxiv. In his medical textbook, William Cheselden had presented "the parts of an hermaphrodite negro, which was neither sex perfect, but a wonderful mixture of both." John B. Davidge questioned Cheselden's categorization of this person as a hermaphrodite and agreed with Parsons that "a real hermaphrodite must possess every sexual part of the male and the female" (168). See William Cheselden, *The Anatomy of the Human Body* (Boston: Manning and Loring, 1795), 314; and John B. Davidge, "Account of the Dissection of a Singular Lusus Nature," *Philadelphia Medical Museum* 2 (1806): 164–69.

32. I use male pronouns when discussing Thomas Hall's life as a man and female pronouns when Hall lived as a woman. When the sources are unclear, I use neutral language.

33. McIlwaine, *Minutes of the Council and General Court of Colonial Virginia, 1622–1632*, 195.

34. Ibid.

35. Mary Beth Norton argues as well that the verdict was unprecedented, and that Hall was probably lonely and "perhaps the target of insults or assaults." See Norton, *Founding Mothers and Fathers*, 196.

36. Alfred F. Young, *Masquerade: The Life and Times of Deborah Sampson, Continental Soldier* (New York: Alfred A. Knopf, 2004), 9.

37. *Pennsylvania Gazette*, August 9, 1764.

38. The first account had said Lewis was thirty-two years old in 1764; the information from the 1770 story would make her seventeen years old in 1764.

39. My thanks to Al Young for sharing his findings on Lewis with me. One of Young's assistants, Paul Uek, tracked down Lewis's genealogical information. If Lewis was born in 1747 (as the 1770 story suggested) and died in 1823 (per the obituary), he lived almost half of his life as a woman. See Charles Edward Banks, M.D., *The History of Martha's Vineyard, Dukes County Massachusetts in Three Volumes* (Edgartown: Dukes County Historical Society, 1966), 3:235.

40. Records of the Middlesex County Court, 1691/92, vols. 1689–99, n.p., as cited in Lawrence W. Towner, "The Indentures of Boston's Poor Apprentices: 1734–1805," *Transactions*, Colonial Society of Massachusetts, 43 (1956–63): 417–68.

41. Otis Hammond, ed., *New Hampshire Court Records, 1640–1692*, New Hampshire State Papers Series, 40 (1943): 96.

42. George F. Dow, ed., *Records and Files of the Quarterly Courts of Essex County, Massachusetts* (Salem, Mass., 1911–21), VI, 341. As cited in Mary Beth Norton, "Communal Definitions of Gendered Identity in Seventeenth-Century English America," in *Through a Glass Darkly: Reflections on Personal Identity in Early America*, ed. Ronald Hoffman, Mechal Sobel, and Fredrika J. Teute (Chapel Hill: University of North Carolina Press, 1997), 53.

43. Susan Juster, " 'Neither Male nor Female': Jemima Wilkinson and the Politics of Gender in Post-Revolutionary America," in *Possible Pasts: Becoming Colonial in Early America*, ed. Robert Blair St. George (Ithaca, N.Y.: Cornell University Press, 2000), 357–79.

44. The 1656 New Haven statute was an exception; it read: "If any man lyeth with mankind, as a man lyeth with a woman, both of them have committed abomination, they both shall surely be put to death. Lev. 20:13. And if any woman change the natural use, into that which is against nature, as Rom. 1:26 she shall be liable to the same sentence, and punishment." See *New Haven's Settling in New England and Some Lawes for Government* (London: printed by M. S. for Livewell Chapman, 1656), in *The True-Blue Laws of Connecticut and New Haven*, ed. J. Hammond Trumbull (Hartford, Conn: American Publishing, 1876), 198–201.

45. Norton, *Founding Mothers and Fathers*, 193.

46. McIlwaine, *Minutes of the Council and General Court of Colonial Virginia, 1622–1632*, 195.

47. His skepticism notwithstanding, Parsons's book is invaluable for the brief vignettes of those "mistakenly" called hermaphrodites. Because Parsons was convinced that hermaphrodites were really "normal" men or women, he urged readers to recognize and avoid past injustices: "Thus it often fared with our reputed Hermaphrodites, who have been banished, tormented, abused, and employed in such Offices as were in themselves severe; cut off from the common Privileges and Freedoms enjoyed by the Publick wheresoever they have been; yea, and put to Death in an inhuman and pitiless Manner" (*Mechanical and Critical Enquiry,* lii).

48. On the role of medical men in ascertaining the cause of marital problems such as impotency that might have been due to intersex conditions, see Michael R. McVaugh, *Medicine before the Plague: Practitioners and Their Patients in the Crown of Aragon, 1285–1345* (Cambridge: Cambridge University Press, 1993), 200–207.

49. Parsons, *Mechanical and Critical Enquiry,* 9. On the "rediscovery" of the clitoris in sixteenth-century Europe, see Katharine Park, "The Rediscovery of the Clitoris," in *The Body in Parts: Fantasies of Corporeality in Early Modern Europe,* ed. David Hillman and Carla Mazzio (New York: Routledge, 1997), 171–93.

50. Nicholas Culpeper, *The Compleat Practice of Physick* (London, 1655), 503. On the relationship between conception and orgasm for women, see Thomas Foster, "Deficient Husbands: Manhood, Sexual Incapacity, and Male Marital Sexuality in Seventeenth-Century New England," *William and Mary Quarterly,* 3rd ser., 56, no. 4 (October 1999): 723–44. Angus McLaren argues that though linking conception and female orgasm lingered into the eighteenth century in England, the notion was relegated to popular culture status as newer ideas of scientific embryology emerged. See McLaren, *Reproductive Rituals: The Perception of Fertility in England from the Sixteenth Century to the Nineteenth Century* (London: Methuen, 1984), 22.

51. Henry Bracken, *The Midwife's Companion; or, a Treatise of Midwifery: Wherein the Whole Art is Explained* (London, 1737), 10.

52. See Dr. Alexander Hamilton, *Outlines of the Theory and Practice of Midwifery* (Philadelphia: Thomas Dobson, 1790), 44–45; William Smellie, *"An Abridgement of the Practice of Midwifery* (Boston: John Norman, 1786), 7, 24; Cheselden, *Anatomy of the Human Body,* 272–73. See also Benjamin Bell, *A System of Surgery* (Worcester, Mass.: Isaiah Thomas, 1791), 367; Andrew Fyfe, *A Compendious System of Anatomy. In Six Parts* (Philadelphia: Thomas Dobson, 1790), 76.

53. Thomas Denman, *An Introduction to the Practice of Midwifery* (New

York: printed by James Oram for William Falconer and Evert Duyckinck, 1802), 34.

54. Parsons, *Mechanical and Critical Enquiry*, 10.

55. As quoted in Park, "Rediscovery of the Clitoris," 178.

56. Sharp, *Midwives Book*, 40. Like Culpeper, Sharp highlighted the clitoris's significance: "The Clitoris will stand and fall as the yard doth, and makes women lustfull and take delight in Copulation, and were it not for this they would have no desire nor delight, nor would they ever conceive" (39).

57. Ibid., 40.

58. Giles Jacob, *A Treatise of Hermaphrodites* (London: E. Curll, 1718), 41–42.

59. Nicholas Venette, *Conjugal Love; or, The Pleasures of the Marital Bed Considered in Several Lectures on Human Generation* (London, 1750). According to Thomas Laqueur, the 1750 edition was the twentieth edition, and there were at least eight French editions before Venette's death in 1698. See Laqueur, *Making Sex: Body and Gender from the Greeks to Freud* (Cambridge, Mass.: Harvard University Press, 1990), 245n5.

60. Jacob, *A Treatise of Hermaphrodites*, 6–8. Stories of women suddenly becoming men had been circulating since at least the sixteenth century. Michel de Montaigne recounted, "Passing through Vitry-le-Francois, I might have seen a man whom the bishop of Soissons had named Germain at confirmation, but whom all the inhabitants of that place had seen and known as a girl named Marie until the age of twenty-two. He was now heavily bearded, and old, and not married. Straining himself in some ways in jumping, he says, his masculine organs came forth; and among the girls there a song is still current by which they warn each other not to take big strides for fear of becoming boys, like Marie Germain." See Michel de Montaigne, *The Complete Essays of Montaigne*, trans. Donald M. Frame (Stanford, Calif.: Stanford University Press, 1958), 69.

61. Jacob, *Treatise of Hermaphrodites*, 16.

62. The term "lesbian" was not used until the late nineteenth century, and so I use it here as merely a shorthand indicating sexual relations between women.

63. Jacob, *Treatise of Hermaphrodites*, 19.

64. Dr. William Handy, "Account of an Hermaphrodite," *Medical Repository of Original Essays and Intelligence* 12 (May-July 1808), 86–87, quotation on 86.

65. Handy wrote, "During copulation, the penis becomes erect, and there is in the moment of ecstacy, an agreeable sensation in all the male parts of generation, but more particularly felt in the glans. There has never existed an inclination for commerce with the female, under any circumstances of excitement of the venereal passion." See Handy, "Account of an Hermaphrodite," 86.

66. Beck and Beck, *Elements of Medical Jurisprudence,* 122

67. Handy, "Account of an Hermaphrodite," 86.

68. Sander L. Gilman, *Difference and Pathology: Stereotypes of Sexuality, Race, and Madness* (Ithaca, N.Y.: Cornell University Press, 1985), esp. 76–108.

69. Park, "Rediscovery of the Clitoris," 171–72.

70. Sharp, *Midwives Book,* 40.

71. Parsons, *Mechanical and Critical Enquiry,* liv.

72. Ibid., 10–11.

73. Carol Groneman, *Nymphomania: A History* (New York: W. W. Norton, 2000). See, for example, Pierre Lefort, M.D, "A Case of Excision of the Clitoris and Lubia Pudendorum," *Medical Repository of Original Essays and Intelligence* 19 (1818): 84–87; Anon., "Extirpation of the Clitoris," *American Medical Review* 2, no. 1 (September 1825), 188.

74. In the eighteenth century, doctors advocated delaying surgery for hypospadias, as long as urine flowed, so that the opening they created would remain open with a catheter. See Nicholas B. Waters, *A System of Surgery Extracted from the Works of Benjamin Bell of Edinburgh* (Philadelphia: T. Dobson, 1791), 166.

75. In a sarcastic newspaper article decrying all kinds of untoward social behavior, one author wrote, "It is wonderful that men, forgetting the dignity of their station on earth, should degenerate into effeminacy, and assume the manners, appearance, and gait of hermaphrodites!" See Anon., *Massachusetts Spy; or, Worcester Gazette* 14 (August 26, 1784), 2.

Chapter Two: From Monsters to Deceivers in the Early Nineteenth Century

Epigraph. James Akin, *Facts Connected with the Life of James Carey, Whose Eccentrick Habits Caused a Post Mortem Examination by the Gentlemen of the Faculty; to Determine Whether He Was Hermaphroditic* (Philadelphia, 1839), frontispiece.

1. John P. Mettauer, M.D., "Practical Observations on those Malformations of the Male Urethra and Penis, termed Hypospadias and Epispadias, with an Anomalous Case," *American Journal of the Medical Sciences* 4 (July 1842): 43–57.

2. Akin, *Facts Connected with the Life of James Carey,* 3. On the postmortem examination of unusual human bodies, including hermaphrodites, for anatomical instruction, see Robert J. Moes and C. D. O'Malley, "Realdo Colombo: 'On Those Things Rarely Found in Anatomy': An Annotated Translation from 'De Re Anatomica' (1559)," *Bulletin of the History of Medicine* 34 (1960), 508–28.

3. Akin, *Facts Connected with the Life of James Carey,* 2. Mention of an odor associated with genital malformations appeared in other cases as well. See

Henry W. Ducachet, "Case of Extraordinary Maleformation of the Genital Organs," *American Medical Recorder* 3 (October 1820): 515–16.

4. Akin, *Facts Connected with the Life of James Carey,* 3.

5. See, for example, James Fitzpatrick, "Account of a Case of Monstrosity," *New York Medical and Physical Journal* 5 (April-June 1826): 317–19. This case described a child's frontal-lobe malformation, which the doctor linked to maternal imagination, if only by discrediting the theory as superstition. In the 1890s, the theory was still circulating. See Edward P. Davis, "Maternal Impression Followed by the Production of a Monster," *American Journal of Obstetrics* 102, no. 4 (1891): 434.

6. Akin, *Facts Connected with the Life of James Carey,* 3, 4.

7. Ibid., 3, 8.

8. Ibid., 8.

9. Perhaps one of the reasons Akin included the minister's narrative of his attempts to "save" Carey is that he was writing a popular pamphlet, not a medical/scientific account. Anything likely to interest a popular audience might make its way into the pamphlet. And just as the public could be titillated and horrified by the descriptions and images of Carey's unusual body, its members might be moved by reports of his spiritual life—a topic of wide interest in the heyday of revivalistic religion. Carey's singularity made him an appealing topic for sensational quasi-journalistic writing.

10. Akin, *Facts Connected with the Life of James Carey,* 5.

11. On the professionalization of medicine, see Paul Starr, *The Social Transformation of American Medicine: The Rise of a Sovereign Profession and the Making of a Vast Industry* (New York: Basic Books, 1982); James C. Mohr, *Abortion in America: The Origins and Evolutions of National Policy, 1800–1900* (New York: Oxford University Press, 1978); Mohr, *Doctors and the Law: Medical Jurisprudence in Nineteenth-Century America* (New York: Oxford University Press, 1993); John Harley Warner, *Against the Spirit of System: The French Impulse in Nineteenth-Century American Medicine* (Princeton, N.J.: Princeton University Press, 1998); and Russell Maulitz, *Morbid Appearances: The Anatomy of Pathology in the Early Nineteenth Century* (Cambridge: Cambridge University Press, 1987). On the ways in which the study of anatomy transformed the field, see Michael Sappol, *A Traffic of Dead Bodies: Anatomy and Embodied Social Identity in Nineteenth-Century America* (Princeton, N.J.: Princeton University Press, 2002).

12. See http://indexcat/nih/nlm/gov/ for the online search engine of the Surgeon-General's Catalogue Index.

13. Karen Halttunen, *Confidence Men and Painted Women: A Study of Middle-Class Culture in America, 1830–1870* (New Haven, Conn.: Yale Univer-

sity Press, 1982); James W. Cook, *The Arts of Deception: Playing with Fraud in the Age of Barnum* (Cambridge, Mass.: Harvard University Press, 2001). Concerns over cross-dressing can be seen in newspaper stories about arrests, even those involving festive burlesques rather than concerted efforts to pass as the opposite sex. In 1847, the Philadelphia *Public Ledger* reported sarcastically, for example: "Three nice young men put forth with some twenty or thirty of their jolly companions to have a grand promenade. Their habits were in such bad taste they were caught foul and with all the trappings, flounces, bustles and all, politely gallanted to the watch house." After imposing an unusually steep fine of $300 each, the judge said, "Nothing is more offensive in the eye of the law . . . than the assumption of that which by nature and art we are not, and cannot be." As quoted in Susan G. Davis, *Parades and Power: Street Theatre in Nineteenth-Century Philadelphia* (Philadelphia: Temple University Press, 1986), 106.

14. A. F[lint], "Hermaphroditism," *Boston Medical and Surgical Journal,* October 7, 1840, 145–47.

15. Ibid., 146.

16. Ibid. We do not know where or when this person was arrested. Laws against cross-dressing multiplied in the mid-nineteenth century. Beginning in the 1840s, cities of every size and in every region of the country enacted gender-normative rules regarding behavior. A law regulating cross-dressing in New York City was first imposed in 1845. Nan Hunter has suggested that these laws focused on gender fraud and targeted women who sought male advantages, such as employment. The New York law made it a crime to assemble "disguised" in public places, though later the law was amended to allow for masquerade balls. See William N. Eskridge Jr., *Gaylaw: Challenging the Apartheid of the Closet* (Cambridge, Mass.: Harvard University Press, 1999), 24–30 and 338–41. Of course biblical injunctions against cross-dressing preceded and justified civil law.

17. F[lint], "Hermaphroditism," 146.

18. Ibid., 146. This recalls the response that Hall provided authorities in seventeenth-century Virginia that s/he had "a peece of fleshe growing at the . . . belly as bigg as the top of his little finger." We don't know if the Hall story had circulated and become urban legend or if the similarity is coincidental. See H. R. McIllwaine, *Minutes of the Council and General Court of Colonial Virginia, 1622–1632* (1670–76; Richmond, Va.: Colonial Press/Everett Waddey, 1924), 194–95.

19. This case bears many similarities to the seventeenth-century Hall case. Both subjects were raised female; both worked at traditionally female occupations; both referred to "a piece of flesh"; and both endured invasive genital scrutiny from "experts."

20. F[lint], "Hermaphroditism," 146.

21. For a fascinating discussion of the link between racial "otherness" and

monstrosity, see Linda Barnes, *Needles, Herbs, Gods, and Ghosts: China, Healing, and the West to 1848* (Cambridge, Mass.: Harvard University Press, 2005).

22. Charles Drake, "A Case of Malformation of the Urinary and Genital Organs," *New York Medical and Physical Journal* 5, no. 3 (July-September 1826): 443–48.

23. *Boston Medical and Surgical Journal* 2 (December 22, 1829): 717–18, quotation on 718.

24. John North, "A Lecture on Monstrosities," *American Medical Intelligencer,* August 15, 1840, 156.

25. Wm. James Barry, "Case of Doubtful Sex," *Medical Examiner, and Record of Medical Science* 10 (May 1847), 308–9, quotation on 308. On nineteenth-century property and registry requirements in Connecticut, see Chilton Williamson, *American Suffrage: From Property to Democracy, 1760–1860* (Princeton, N.J.: Princeton University Press, 1960), 276–79; and Marchette Chute, *The First Liberty: A History of the Right to Vote in America, 1619–1850* (New York: Dutton, 1969), 311–13.

26. Barry, "Case of Doubtful Sex," 308.

27. Ibid., 308–9, quotation on 309.

28. Ibid., 309, 308. This phrase recalls Samuel Farr's definition of a hermaphrodite as one "partaking of the distinguishing marks of both sexes." See Farr, "Elements of Medical Jurisprudence" (London, 1788), reprinted in Thomas Cooper, ed., *Tracts of Medical Jurisprudence* (Philadelphia, 1819), 13.

29. Daniel A. Cohen, ed., *The Female Marine and Related Works: Narratives of Cross-Dressing and Urban Vice in America's Early Republic* (Amherst: University of Massachusetts Press 1997). Other stories of deception circulated in popular literature. See, for example, Anon., "A Male Abbess," *Harper's Weekly,* June 12, 1858, 375.

30. "Hermaphroditism," *Boston Medical and Surgical Journal,* February 13, 1850, 45–46; "The Clitoris," *Medical Examiner,* May 18, 1839, 314–15.

31. Benjamin Smith Barton, "Account of Henry Moss, A White Negro," *Philadelphia Medical and Physical Journal* 2 (part II, 1806): 3–18; Charles W. Peale, "Account of a Negro, or a Very Dark Mulatto, Turning White," *Massachusetts Magazine; or, Monthly Museum* 3 (December 1791), 744. The preoccupation with changing race persisted into the twenty-first century; the play *White Chocolate,* by William Hamilton, which portrayed a New York couple who transform from white to black overnight, opened in New York City on October 6, 2004. *New York Times,* October 9, 2004, B18.

32. Anon., "A Question of Legitimacy," *Western Journal of Medicine and Surgery* 3, no. 5 (May 1845): 457–59, quotations on 457.

33. Ibid., 457.

34. Ibid., 458.

35. Felix Pascalis, "Desultory Remarks on the Cause and Nature of the Black Colour in the Human Species; Occasioned by the Case of a White Woman Suddenly Turned Black," *Medical Repository of Original Essays and Intelligence* 19, no. 4 (1818): 366–71, quotation on 368. On anxieties about race in this period, see Barnes, *Needles, Herbs, Gods, and Ghosts;* Joanne Pope Melish, *Disowning Slavery: Gradual Emancipation and Race in New England, 1780–1860* (Ithaca, N.Y.: Cornell University Press, 1998), esp. 141–62; and Martha Hodes, "The Mercurial Nature and Abiding Power of Race: A Transnational Family Story," *American Historical Review* 108 (February 2003): 84–119. See also Gary Nash, *Race and Revolution* (Madison, Wisc.: Madison House, 1990), 81–83; and James Forten, *Letters from a Man of Colour on a Late Bill Before the Senate of Pennsylvania, 1813*, in Nash, *Race and Revolution*, 190–98.

36. W. S. Forwood, "The Negro—A Distinct Species," *Medical and Surgical Reporter* 10, no. 5 (May 1857): 225–35; Forwood, "The Negro—A Distinct Species, No. 2," *Medical and Surgical Reporter* 11, no. 2 (February 1858): 69–96, quotation on 71.

37. Senex, "Is the Negro a Distinct Species? Answered in the Negative," *Medical and Surgical Reporter* 10, no. 6 (June 1857): 288–302, quotation on 294. A reprint of Smith's essay can be found in Samuel Stanhope Smith, *An Essay on the Causes of the Variety of Complexion and Figure in the Human Species* (1787; rev. ed. 1810; repr. Cambridge, Mass.: Harvard University Press, 1965). See also Elihu, "Some Remarks in Regard to the Ethnological Discussion in Progress in the Pages of the *Reporter*," *Medical and Surgical Reporter* 10, no. 8 (August 1857): 389–92; and John Ewcorstart, "The Negro not a Distinct Species," *Medical and Surgical Reporter* 10, no. 12 (December 1857): 577–83.

38. Forwood, "The Negro—A Distinct Species, No. 2," 77.

39. As quoted in ibid., 79. According to H. Hotz, the translator of Count Arthur de Gobineau's 1856 *The Moral and Intellectual Diversity of Races* (Philadelphia: Lippincott, 1856): "If there are, or ever have been, external agencies that could change a white man into a negro, or vice versa, it is obvious that such causes have either ceased to operate, or operate only in a lapse of time so incommensurable as to be imponderable to our perceptions" (79).

40. On the manifestation of medical disorders in blacks, see Harriet A. Washington, *Medical Apartheid: The Dark History of Medical Experimentation on Black Americans from Colonial Times to the Present* (New York: Doubleday, 2006).

41. Frederick Hollick, *The Marriage Guide; or, Natural History of Generation: A Private Instructor for Married Persons and Those About to Marry* (New York: American News Company, 1850), 140.

42. Francis Wharton and Moreton Stillé, *A Treatise on Medical Jurisprudence* (Philadelphia: Kay and Bro., 1855), 312. This case was taken from an earlier description by Everard Home concerning a woman of the Mandingo nation, purchased in the West Indies in 1774. See Everard Home, "The Dissection of an Hermaphrodite Dog. With Observations on Hermaphrodites in General," *Philosophical Transactions of the Royal Society of London,* abridged 18 from 1796–1800 (1809), 485–96, quotation on 488. For other examples of the conflation between medical anomalies and African Americans, see William E. Moseley and Robert B. Morison, "Elephantiasis Arabum of the External Genitals of a Negress," *Medical News* 50 (April 1887): 462–63; Dr. Cartwright, "Diseases and Peculiarities of the Negro Race," DeBow's Review of the Southern and Western States 1, no. 1 (July 1851): 64–69; Anon. [by a Professional Planter], *Practical Rules for the Management and Medical Treatment of Negro Slaves in the Sugar Colonies* (London: J. Barfield, 1811), esp. 371.

43. For other penile anomalies in African Americans, see, for example, C. B. Barrett, "Fibrous Induration of the Prepuce and Body of the Penis—Amputation," *New York Journal of Medicine and Collateral Sciences* 4, no. 12 (May 1845): 358–60, esp. 358. On the amputation of an African American man's penis, see Thomas L. Ogier, "Case of Induration and Enlargement of the Body of the Penis, with a New Method of Amputating that Organ," *American Journal of the Medical Sciences* 36 (August 1836), 382–86.

44. J. W. Heustis, "Malformation of the Sexual Organs," *The American Journal of the Medical Sciences* 14 (February 1831): 555–63, quotation on 557.

45. S. B. Harris, "Case of Doubtful Sex," *American Journal of the Medical Sciences* 14, no. 27 (July 1847): 121–23, quotations on 123.

46. For early American efforts to establish a biological basis for "racial difference," see Samuel George Morton, *Crania Americana; or, a Comparative View of the Skulls of Various Aboriginal Nations of North and South America, to which is Prefixed an Essay on the Varieties of the Human Species* (Philadelphia: Simpkin, Marshall, 1839). On biological divisions and scientific racism, see Nancy Leys Stepan, "Race, Gender, Science and Citizenship," *Gender and History* 10, no. 1 (April 1998): 26–52.

47. In his account of a hermaphrodite orangutan, Richard Harlan made the point that a perfect set of male and female organs could be found in even the highest class of animals, primates. He also described a person living in Lisbon whose genitals "united both sexes in apparently great perfection." See Richard Harlan, *Description of an Hermaphrodite Orang Outang. Lately Living in Philadelphia* (Philadelphia, 1827).

48. Farr, "Elements of Medical Jurisprudence," 1–80, quotation on 13; James Parsons, *Mechanical and Critical Enquiry into the Nature of Hermaphrodites*

(London: J. Walthoe, 1741), 2. On the issue of self-impregnation, see, for example, Francis Wharton and Moreton Stillé, *A Treatise on Medical Jurisprudence* (Philadelphia: Kay and Bro., 1855), 311; and Louis Agassiz and A. A. Goulde, "Generation—Reproduction," *Southern Quarterly Review,* 9 (April 1854): 333–46. In the twentieth century doctors occasionally mentioned the possibility, if only to deny it. See Boris Kwartin and Joseph Hyams, "True Hermaphroditism in Man," *Journal of Urology* 18 (1927): 363–83; H. F. Brewster and H. E. Cannon, "Hermaphroditism; Report of Case of Pseudo-Hermaphroditism," *New Orleans Medical and Surgical Journal* 82 (1929–30): 76–80.

49. Amos Dean called Geoffroy Saint-Hilaire's book "an ingenius work on hermaphroditism." See Dean, *Principles of Medical Jurisprudence: Designed for the Professions of Law and Medicine* (Albany, N.Y., 1850), 20. On the accessibility of Dean's book, see Mohr, *Doctors and the Law,* 37.

50. Isidore Geoffroy Saint-Hilaire quoted in Beck and Beck, *Elements of Medical Jurisprudence,* 132; North, "Lecture on Monstrosities," 156. Amos Dean agreed that hermaphrodites were impossible; far from being able to function sexually as men and women, "there exists the incapacity of exercising the functions of either." See Dean, *Principles of Medical Jurisprudence,* 17.

51. John Neill, "Case of Hermaphroditism," *Summary of the Transactions of the College of Physicians of Philadelphia* 1 (1850): 113–15, quotations on 113 and 114.

52. Ibid., 115.

53. Wharton and Stillé, *Treatise on Medical Jurisprudence,* 313.

54. Neill, "Case of Hermaphroditism," 113.

55. For an example of a case in which the person's hermaphroditism was discovered only at his autopsy, see Anon., "Hermaphrodism," *American Journal of the Medical Sciences* 13 (1833): 222.

56. Wharton and Stillé, *Treatise on Medical Jurisprudence,* 311–12.

57. Warren was an influential surgeon at Harvard and one of the founders of Massachusetts General Hospital in Boston.

58. See John C. Warren, "Non-existence of Vagina, Remedied by an Operation," *American Journal of the Medical Sciences* 13 (November 1833): 79–80. Emphasis added.

59. S. D. Gross, "Case of Hermaphrodism Involving the Operation of Castration and Illustrating a New Principle in Juridical Medicine," *American Journal of the Medical Sciences* 48 (October 1852): 386–90, quotations on 387. Christina Matta has also analyzed this case, and we have arrived at similar conclusions independently. See Matta, "Ambiguous Bodies and Deviant Sexualities: Hermaph-

rodites, Homosexuality, and Surgery in the United States, 1850–1904," *Perspectives in Biology and Medicine* 48 (Winter 2005): 74–83.

60. Gross, "Case of Hermaphrodism," 387–88.

61. Ibid., 388, 390. Gross may have operated to satisfy the parents' wishes, but the justification he provided in the case report emphasized the girl's marriageability.

62. Ibid., 387, 389, 388.

63. Wharton and Stillé, *Treatise on Medical Jurisprudence,* 317.

64. George Blackman, "On Hermaphroditism; with an Account of Two Remarkable Cases," *American Journal of the Medical Sciences* 51 (July 1853): 63–68, quotation on 63. On early gynecological surgery, see Deborah Kuhn McGregor, *From Midwives to Medicine: The Birth of American Gynecology* (New Brunswick, N.J.: Rutgers University Press, 1998). On the history of anesthesia, see Martin S. Pernick, *A Calculus of Suffering: Pain, Professionalism, and Anesthesia in Nineteenth-Century America* (New York: Columbia University Press, 1985).

65. T. Holmes, *A System of Surgery, Theoretical and Practical, in Treatises by Various Authors* (London: Longman, Green, Longman, Roberts, and Green, 1864), 820.

66. Lawson Tait, "Hermaphroditism," *Transactions of the American Gynecological Society for the Year 1876* 1 (1877): 318–25, quotation on 318–19.

67. Hollick, *Marriage Guide,* 292.

68. Ibid., 292.

69. Ibid., 293.

70. See, for example, Horatio R. Bigelow, "The History of Hermaphroditism," *New England Medical Monthly* 4 (October 1884–85): 1–7. American doctors shared the confusion of criteria with their European colleagues, yet all were determined to fit people into two distinct sexes, despite contradictory evidence. See Alice Domurat Dreger, *Hermaphrodites and the Medical Invention of Sex* (Cambridge, Mass.: Harvard University Press, 1998), esp. 83–109. On the developing notion of a "true sex," see also Michel Foucault, "Introduction" to *Herculine Barbin,* by Herculine Barbin, trans. Richard McDougal (New York: Pantheon, 1980), vii–xvii.

71. Beck and Beck, *Elements of Medical Jurisprudence,* 129. In the 1890s, a Dr. Tillotson worried publicly about the consequences (for doctors more than patients) of a wrong decision: "Should our decision chance to be wrong, in after years it may humiliate the family and the person, and ruin the reputation of the physician in that community; for what woman would employ an obstetrician who would be guilty of incorrectly diagnosticating [*sic*] the sex of her child?" See

D. J. Tillotson, "Case of a Hermaphrodite," *Medical and Surgical Reporter* 63, no. 23 (December 1890): 647.

72. E. Noeggerath, "The Diseases of Blood-Vessels of the Ovary in Relation to the Genesis of Ovarian Cysts," *American Journal of Obstetrics* 13 (January 1880): 174–76, quotation on 176. Unlike most doctors, E. Noeggerath believed that hermaphrodites existed in the human species, but he defined the term more broadly than most, as "an animal in which there exists a mixture of the male and female organs" (175).

73. Medical examinations for military service often revealed hermaphroditic conditions. One man, described by the doctor as "a very interesting specimen," was discovered in an exam for recruits for the United States Naval Service. He had served for ten to twelve years, even though he exhibited "in every way the appearance of a well-developed female." In an unusual eroticized account, the doctor noted "His breasts are full, round, soft, and beautiful, with well-developed nipple and areolae, and identical with the breast of a virgin. The curves of the thigh, so characteristic of the female, are well shown upon this truly wonderful form." See J. W. Bragg, "A Case of Hermaphroditism," *Boston Medical and Surgical Journal* 65 (1862): 349–50.

74. [B. Cloak], "A Case of Hermaphroditism," *Medical and Surgical Reporter* (September 10, 1864): 71–73, quotation on 72.

75. Ibid., 72.

76. Ibid., emphasis added.

77. Ibid., 73. Beck and Beck recounted a similar case in 1834 that highlighted sexual desire in the medical decision to assign gender. The person had undeveloped breasts and the voice and mustache of a man, as well as "strong sexual desires." After thirty-three years of living as a woman, "he" was told by doctors that he was actually a man, and he "then assumed the male attire." See Beck and Beck, *Elements of Medical Jurisprudence,* 120.

78. Dreger, *Hermaphrodites.*

Chapter Three: The Conflation of Hermaphrodites and Sexual Perverts at the Turn of the Century

Epigraph. Anon., "Hermaphrodism," *Medical Record* 41, no. 5 (January 1892): 115.

1. "Correction" may have been the goal, but as many intersex people have noted, surgical solutions sometimes brought more harm than good, including further surgery to eliminate scar tissue and procedures that resulted in the loss of sexual sensation. See Howard Devore, "Growing up in the Surgical Maelstrom,"

and Sven Nicholson, "Take Charge!: A Guide to Home Catheterization," both in *Intersex in the Age of Ethics,* ed. Alice Domurat Dreger (Hagerstown, Md.: University Publishing Group, 1999), 78–81, 201–8.

2. Of course, whatever is conventional at a certain time and place will be "imposed" on child patients; children, let alone infants, are not given free rein to make medical decisions.

3. Heidi Walcutt, "Time for a Change," in Dreger, *Intersex in the Age of Ethics,* 197–200, quotation on 198.

4. A few physicians considered hermaphrodites to be neuter and so did not believe they should marry at all. See Anon., "Hermaphrodism," 115.

5. For interpretations of these studies, see Richard C. Pillard, "The Search for a Genetic Influence on Sexual Orientation," and Garland E. Allen, "The Double-Edged Sword of Genetic Determinism: Social and Political Agendas in Genetic Studies of Homosexuality, 1940–1944," both in *Science and Homosexualities,* ed. Vernon A. Rosario (New York: Routledge, 1997), 226–41, 243–70.

6. J. W. Underhill, "Two Hermaphrodite Sisters," *American Journal of Obstetrics and Diseases of Women and Children* 13, no. 1 (January 1880): 174–76, quotation on 175. See also Henry Avery, "A Genuine Hermaphrodite," *Medical and Surgical Reporter* 19 (1868): 144. In his medical jurisprudence textbook, Gilbert H. Stewart defined hermaphrodites even more broadly as "all persons afflicted with malformation of the sexual organs." See Stewart, *Legal Medicine* (Indianapolis: Bobbs-Merrill, 1910), 117.

7. Underhill, "Two Hermaphrodite Sisters," 175. Emphasis added.

8. J. W. Long, "Hermaphrodism So-Called, with an Illustrative Case," *International Journal of Surgery* 9, no. 8 (August 1896): 243–44, quotation on 243.

9. Lawson Tait objected to the use of " 'spurious hermaphroditism,' a term, the very composition of which is enough to condemn it." See Tait, "Hermaphroditism," 318–25, quotation on 318. For the view that most so-called hermaphrodites were really men with undescended or atrophied testicles, see Basile Poppesco, *Hermaphrodism, From a Medical-Legal Point of View,* trans. Edward Warren Sawyer (Chicago: W. B. Keen, Cooke and Co., 1875).

10. Dr. Maurice A. Walker wrote, "I will say that cases of true complete hermaphroditism recorded are about as rare as the frequently-mentioned hen's teeth" (435). Walker's patient considered himself male, and had nocturnal emissions and occasional sexual relations with women, but he also bled monthly, through his nose, accompanied by an "ill-defined feeling of tenderness in the loins and lateral pelvic regions" (434). See Walker, "A Case of Pseudo-hermaphroditism," *New York Medical Journal* 60 (1894): 434–35. On the relationship between menstruation and nosebleeds, see Leon J. Saul, "Feminine Significance of the Nose," *Psy-*

choanalytic Quarterly 17 (1948): 51–57; Marc Hollendar, "Observations on Nasal Symptoms: Relationship of the Anatomical Structure of the Nose to Psychological Symptoms," *Psychiatric Quarterly* 30, no. 1 (December 1956): 375–86.

11. G. Frank Lydston, *Impotence and Sterility with Aberrations of the Sexual Function and Sex-Gland Implantation* (Chicago: Riverton Press, 1917), esp. 7–24, quotations on 7–8.

12. Lydston's reference to the subject's race was not unique. Doctors frequently described a patient as nonwhite and mentioned lower-class status, particularly if the person were involved in some sort of dishonest activity.

13. Hypospadias and Epispadias Association, www.heainfo.org; American Urological Association, www.urologyhealth.org/pediatric/index.cfm?cat=01& topic=31. See, for example, Guy Hinsdale, "A Case of Hypospadias with Hermaphroditic Appearance," *Transactions of the College of Physicians of Philadelphia* 17 (1895): 18–19.

14. Lydston, *Impotence and Sterility*, 9. According to Dr. John Gatti, professor of pediatric urology at Mercy Children's Hospital and Kansas University Medical Center, "The combination of hypospadias and undescended testis can be an indicator of an underlying intersex disorder. In a 1999 study by Kaefer et al., intersex states were identified in approximately 30% of patients with unilateral or bilateral undescended testes and hypospadias, and more proximal meatal location carried a higher association with intersex states than more distal meatal location. If any gonad was nonpalpable, the incidence rose to 50%; however, if both gonads were palpable, the incidence was only 15%" (John M. Gatti, Andrew J. Kirsch, and Howard M. Snyder III, "Hypospadias," *eMedicine: The Medscape Journal;* www .emedicine.com/ped/topic1136.htm). Perhaps the cook was bisexual; it is not clear whether or not Lydston ever examined the patient. Rather than describing physical conformation, Lydston conjectured instead about the cook's possible sexual versatility.

15. George DuBois Parmly, "Hermaphrodism," *American Journal of Obstetrics* 19 (1886): 931–46, quotation on 933. See also Sir James Y. Simpson, *Anaesthesia, Hospitalism, Hermaphroditism and a Proposal to Stamp out Small-Pox and Other Contagious Diseases* (Edinburgh: Adam and Charles Black, 1871), esp. 410–19. The *New York Times* reported about a person who "passed" as a man for sixty years and married a woman, originally adopting men's dress because "when a young woman she found it difficult to make her way on account of her sex." Anon., "Passed Sixty Years as a Man," *New York Times,* November 12, 1907, 1.

16. Parmly, "Hermaphrodism," 934.

17. Ibid.

18. Ibid., 933–34. Emphasis added.

19. Ibid., 934–36, quotation on 936. The tendency to unite hermaphroditism with homosexuality had legal ramifications as well. Douglas C. Baynton describes several instances of émigrés being deported in the early twentieth century for various disabilities, including that of a "supposed woman" who was refused entry to the United States in 1908 because she was considered a hermaphrodite. Officials justified her exclusion on the grounds that hermaphrodites were "usually of perverted social instincts and with lack of moral responsibility (38)." Even though her family committed to her financial support, she was denied admission. See Baynton, "Defectives in the Land: Disability and American Immigration Policy, 1882–1924" *Journal of American Ethnic History* (Spring 2005): 31–44.

20. Jennifer Terry, *An American Obsession: Science, Medicine, and Homosexuality in Modern Society* (Chicago: University of Chicago Press, 1999), 50–53.

21. G. Frank Lydston, "Sexual Perversion, Satyriasis and Nymphomania," *Medical and Surgical Reporter* 61, no. 10 (September 7, 1889): 253–58. See also Lydston, *Impotence and Sterility*, 10–14. For a later discussion of acquired and congenital inversion, see J. Herbert Claiborne, "Hypertrichosis in Women: Its Relation to Bisexuality (Hermaphroditism)," *New York Medical Journal* 99 (1914): 1178–84.

22. Lydston, *Impotence and Sterility*, 14. Lydston wrote, "It is not only charity, but a sense of justice and a desire to lessen the stigma upon human nature, that impels the author to include typic cases of sexual perversion under the head of aberrant sexual differentiation, and to attribute the condition to perverted or imperfect evolutionary development, on the one hand, and a reversion of type, on the other" (19).

23. See Bransford Lewis, "A Case of Hermaphrodism(?)" *Medicine* 2, no. 10 (October 1896): 793–96 for the use of the word "confusion." Of one patient, Lewis described, "The confusion in her anatomical make-up is readily apparent," and "the genitals appear embarrassingly confused in their anatomy" (794–95). On the relationship between inversion and psychosexual hermaphroditism, see also J. Richardon Parke, *Human Sexuality: A Medico-Literary Treatise on the History and Pathology of the Sex Instinct for the Use of Physicians and Jurists,* 4th ed. (Philadelphia: Professional Publishing, 1909).

24. C. W. Allen, "Report of a Case of Psychosexual Hermaphroditism," *Medical Record*, 51, no. 9 (May 1897): 653–55. This was likely the Florence Crittenton Mission, a home opened by Charles Crittenton in 1883 in New York City to serve "lost and fallen girls," and named in memory of his own daughter who died at age four of tuberculosis. See Regina Kunzel, *Fallen Women, Problem Girls: Un-*

married Mothers and the Professionalization of Social Work, 1890–1945 (New Haven, Conn.: Yale University Press, 1995).

25. C. W. Allen, "Report of a Case of Psychosexual Hermaphroditism," 653.

26. Ibid., 654.

27. Ibid.

28. Ibid. On the relationship between gender inversion and male effeminacy, see George Chauncey, *Gay New York: Gender, Urban Culture, and the Making of the Modern Gay World 1890–1940* (New York: Basic Books, 1994), 47–63.

29. C. W. Allen, "Report of a Case of Psychosexual Hermaphrodism," 655.

30. Joanne Meyerowitz, *How Sex Changed: A History of Transsexuality in the United States* (Cambridge, Mass.: Harvard University Press, 2002), 4–5. Shannon Price Minter has explained how the shift from an emphasis on gender inversion to an emphasis on sexual object choice in gay rights advocacy has, in effect, excluded transgender people from inclusion in the modern gay rights movement. See Minter, "Do Transsexuals Dream of Gay Rights?" in *Transgender Rights*, ed. Paisley Currah, Richard M. Juang, and Shannon Price Minter (Minneapolis: University of Minnesota Press, 2006), 141–70.

31. William Lee Howard, "Psychical Hermaphroditism," *Alienist and Neurologist,* 18, no. 2 (April 1897): 111–15, quotation on 113. Howard's article was summarized and elaborated upon in the *Medical and Surgical Reporter.* The author discussed treatment options for the "true sexual pervert." Drug medication was useless, he argued, but keeping them free of surroundings that would incite emotional feelings, "such as music, art galleries or the theatres" could stem "vicious habits," though they would remain "psychically inverted." See Anon., *Medical and Surgical Reporter* 77, no. 10 (September 1897): 300–301.

32. Howard, "Psychical Hermaphroditism," 115.

33. Interestingly, some contemporary transsexuals are now positing this same theory to explain their desire to live as the other gender. Comparing themselves to intersexed people, they argue that their condition is hidden in the brain, but biological nonetheless. See Alice D. Dreger and April Herndon, "Progress and Politics in the Intersex Rights Movement: Feminist Theory in Action," *GLQ: A Journal of Lesbian and Gay Studies* 15, no. 2 (March 2009): 199–224.

34. See the Androgen Insensitivity Support Group Web site for comprehensive information, including contemporary personal narratives, on this metabolic condition, www.aissg.org. For a similar case, see L. H. Luce and W. H. Luce, "Three Cases of Hypospadias in Which the Sex Was Undeterminable until Puberty," *American Naturalist* 24 (November 1890): 1016–19.

35. J. B. Naylor, "A Case of Hermaphroditism," *Columbus Medical Journal* 17 (1896): 265–68, quotation on 266.

36. Ibid., 267.

37. Anon., "Hermaphroditism, or Sexual Perversion," *Medical and Surgical Reporter* 63, no. 15 (October 11, 1890): 433.

38. Ibid.

39. Long, "Hermaphrodism So-Called," 244.

40. Doctors frequently mentioned that their patients made money exhibiting themselves, sometimes in public venues, and often to other doctors eager to learn from these bodies. See, for example, Henry W. Ducachet, "Case of Extraordinary Maleformation [*sic*] of the Genital Organs," *American Medical Recorder* 3, no. 4 (October 1820): 515–16. Some physicians deplored so-called scientists who would pay to see such exhibitors. See Anon., *Medical News*, 46, no. 10 (March 7, 1885): 268. See also Alice Domurat Dreger, "Jarring Bodies: Thoughts on the Display of Unusual Anatomies," *Perspectives in Biology and Medicine* 43, no. 2 (Winter 2000): 161–72.

41. Samuel E. Woody, "An Hermaphrodite," *Louisville Medical Monthly* 2 (1896): 418–19, quotation on 419.

42. Ibid., 419. Emphasis in original.

43. Anon. "Case of a Hermaphrodite Aged Nine Years, with the External Appearance of a Female, in Whom Both Testicles Were Removed from the Labia Majora," *Medical and Surgical Reporter* 53, no. 1 (July 4, 1885): 19–21, quotation on 20. Emphasis added. For a case in which an older female patient wanted testicles removed, see W. B. Wells, "Report of Surgical Cases," *Transactions of the Annual Session of the Medical Society of the State of Tennessee* (1889), 198–205.

44. See chapter 2 for a discussion of Dr. Samuel Gross's surgical case.

45. Contemporary intersex activists have written extensively on doctors' commitments to cultural norms. See especially Cheryl Chase, " 'Cultural Practice' or 'Reconstructive Surgery'? U.S. Genital Cutting, the Intersex Movement, and Medical Double Standards," in *Genital Cutting and Transnational Sisterhood*, ed. Stanlie M. James and Claire C. Robertson (Urbana: University of Illinois Press, 2002), 126–51. See also Nancy Ehrenreich and Mark Barr, "Intersex Surgery, Female Genital Cutting, and the Selective Condemnation of 'Cultural Practices,' " *Harvard Civil Rights—Civil Liberties Law Review* 40 (2005): 71–140.

46. E. Cross, "Report of Two Cases of Occlusion of the Vagina; One from Traumatism, the Other Congenital," *American Journal of Obstetrics* 19 (1886): 802–3, quotation on 803.

47. See Deborah Kuhn McGregor, *From Midwives to Medicine: The Birth of American Gynecology* (New Brunswick, N.J.: Rutgers University Press, 1998) for a discussion of mortality from gynecological surgery.

48. Thomas Addis Emmet, *The Principles and Practice of Gynaecology* (Philadelphia: Henry C. Lea, 1884), 144–45.

49. George Tully Vaughan, "The Restriction of Surgical Operations," in Vaughan, *Papers on Surgery and Other Subjects* (Washington, D.C.: W. F. Roberts, 1932), 360. This article was originally published in the *Charlotte Medical Journal* in 1908. On the hazards of penile surgery, see Anon., "Hypospadias and Epispadias," *Medical News* (December 2, 1872): 234–38; and W. Stump Forwood, "A Rare Surgical Accident," *Medical and Surgical Reporter* 25, no. 14 (September 1871): 293–95. Some doctors described genital surgery more cavalierly. Referring to the amputation of a clitoris that looked like a penis, gynecological surgeon Lawson Tait remarked in 1876, "A scratch with the surgeon's knife would at once remove all possibility of error." See Tait, "Hermaphroditism," 321.

50. E. Cross, "Report of Two Cases," 803. For a patient who refused similar surgery, see H. Lassing, "A Case of Hermaphroditism," *Medical and Surgical Reporter* 51, no. 19 (November 1884): 516.

51. William P. McGuire, "A Case of Mistaken Sex," *Medical News* 44, no. 7 (February 16, 1884): 186.

52. Ibid.

53. Shelley Tremain, "On the Subject of Impairment," in *Disability/Postmodernity: Embodying Disability Theory*, ed., Mairian Corker and Tom Shakespeare (New York: Continuum, 2002), 32–47. Cheryl Chase has since changed her name to Bo Laurent.

54. Ed., "A Case of Mistaken Sex," *Maryland Medical Journal: A Weekly Journal of Medicine and Surgery* 10, no. 43 (February, 23, 1884): 762–63, quotation on 763.

55. Ibid., 763.

56. William P. McGuire, "A Case of Mistaken Sex," *Maryland Medical Journal: A Weekly Journal of Medicine and Surgery* 10, no. 44 (March 1, 1884): 774; see also Anon., "A Case of Mistaken Sex," *Medical News* 44, no. 10 (March 8, 1884): 291.

57. McGuire, "Case of Mistaken Sex" (March 1, 1884), 774.

58. James L. Little, "Spurious Hermaphroditism. A Case of Hypospadias, Where the Patient, Mistaken for a Female at Birth, Has Passed as Such to the Present Time," *Illustrated Medicine and Surgery* 2 (1882): 180–82, quotation on 182.

59. Ibid., 182.

60. Ibid., 182.

61. Ibid., 181.

62. Underhill, "Two Hermaphrodite Sisters," 176.

63. Little, "Spurious Hermaphroditism," 181.

64. Ibid., 182.

65. J. Riddle Goffe, "A Pseudohermaphrodite, in Which the Female Characteristics Predominated: Operation for Removal of the Penis and the Utilization of the Skin Covering it for the Formation of a Vaginal Canal," *American Journal of Obstetrics and Diseases of Women and Children* 48 (December 1903): 755–63, esp. 759; Fred J. Taussig, "Shall a Pseudohermaphrodite Be Allowed to Decide to Which Sex He or She Shall Belong?" *American Journal of Obstetrics and Diseases of Women and Children* 49 (January 1904): 162–65. See Matta, "Ambiguous Bodies." I am grateful to Christina Matta for sharing this case with me.

66. Goffe, "Pseudohermaphrodite," 757. For a similar case, but one in which the woman showed "no sexual desire and has no voluptuous sensations in the clitoris," see Francis Minot, "A Case of Apparent Hermaphroditism," *Boston Medical and Surgical Journal* 133, no. 5 (August 1895): 112.

67. Perhaps Goffe was unwilling to "unsex" his patient because he was convinced that as a woman her sexual proclivities toward men would be appropriate.

68. J. Riddle Goffe, Fred J. Taussig, and Franz von Neugebauer, "Correspondence: Hermaphroditism and the True Determination of Sex," *Interstate Medical Journal* 11 (May 1904): 314–18, quotations on 314–15. The emphasis on ovarian and testicular tissue follows Alice Dreger's "age of gonads" hypothesis. See Dreger, *Hermaphrodites,* 139–66.

69. Goffe, "Pseudohermaphrodite," 763; Goffe, Taussig, and Neugebauer, "Correspondence," 315. On the controversy, see Geertje Mak, " 'So We Must Go behind Even What the Microscope Can Reveal': The Hermaphrodite's 'Self' in Medical Discourse at the Start of the Twentieth Century," *GLQ: A Journal of Lesbian and Gay Studies* 11, no. 1 (2005): 65–94.

70. James Parsons, *Mechanical and Critical Enquiry into the Nature of Hermaphrodites* (London: J. Walthoe, 1741), xxxiv.

71. William S. Gottheil and Carol Goldenthal, "Pseudohermaphroditism," *New York Medical Journal,* May 19, 1917, 1–11, quotations on 4, 7, 6, 3. Mental illness and hermaphroditism were commonly linked. See, for example, Edward Swasey, "An Interesting Case of Malformation of the Female Sexual Organs," *American Journal of Obstetrics and Diseases of Women and Children* 14, no. 1 (January 1881): 94–195; John McDowell McKinney, "An Unusual Case of Heterogenitalism," *Association for the Study of Internal Secretions* 9 (1925): 229–31; and F. S. White, "A Hermaphrodite (?) in Insane Asylum," *Daniel's Texas Medical Journal* 6 (1890–91): 236–37; Some denied the link between hermaphroditism and mental illness because there were so few hermaphrodites among asylum patients. See H. F. Brewster and H. E. Cannon, "Hermaphroditism; Report of Case of Pseudo-Hermaphroditism," *New Orleans Medical and Surgical Journal* 82 (1929–30), 76–80; J. Allen Jackson, Horace V. Pike, and Ida Ashenhurst,

"Pseudohermaphroditism in a Mental Patient," *Medical Journal and Record* 128 (1928): 232–33.

72. Gottheil and Goldenthal, "Pseudohermaphroditism," 11, 7.

73. The inclination to legitimate one's own point of view by proving earlier doctors wrong also persisted.

Chapter Four: Cutting the Gordian Knot

Epigraph. Anon, "Our Berlin Letter," *Medical News* 72, no. 10 (March 5, 1898): 314.

1. Leon L. Solomon, "Hermaphroditism: Report of a Case of Apparently True Hermaphroditism with Photographs of the 'Woman,'" *Medical Record* 35 (1929), 553–58, quotation on 554.

2. Ibid., 557.

3. From Shakespeare, *King Lear* (3.2,59).

4. Solomon, "Hermaphroditism," 557, 558.

5. W. Blair Bell, *The Sex Complex: A Study of the Relationships of the Internal Secretions to the Female Characteristics and Functions in Health and Disease* (New York: William Wood 1916), 13; Douglas C. McMurtrie, "The Theory of Bisexuality: A Review and Critique," *The Lancet-Clinic* 109 (1913): 370–72. See also J. Herbert Claiborne, "Hypertrichosis in Women: Its Relation to Bisexuality (Hermaphroditism)," *New York Medical Journal* 99 (1914): 1178–84. Claiborne argued that "Man is essentially bisexual and this is evidenced by the existence in the male of the mammary glands and in the female of the clitoris" (1179). In the 1930s Emil Witschi, professor of zoology at the University of Iowa, questioned the bisexuality theory by contending that all true human hermaphrodites were genetically female. See Witschi, "Sex Deviations, Inversions, and Parabiosis," in *Sex and Internal Secretions: A Survey of Recent Research*, ed. Edgar Allen (Baltimore: Williams and Wilkins, 1934), 227.

6. Emil Novak, "Sex Determination, Sex Differentiation, and Intersexuality," *Journal of the American Medical Association* 105 (1935): 413–20, quotation on 413. Some doctors conflated the bisexual germ cell theory with their patient's "psychical hermaphroditism." Frederick L. Patry's patient said that his penis turned into a clitoris at night and that he had both a man and a woman inside of him. Patry agreed that his patient was "psychically part woman and part man, and could have sex relations with himself." He attributed this to the theory of bisexuality, which could manifest solely on the cellular level or "with the psychic or somatic signs of hermaphroditism" (434). Patry, "Theories of Bisexuality with Report of a Case," *Psychoanalytic Review* 15 (1928): 417–39.

7. Joanne Meyerowitz, *How Sex Changed: A History of Transsexuality in the*

United States (Cambridge, Mass.: Harvard University Press), 22–29; James Kiernan, "Bisexuality," *Urologic and Cutaneous Review* 18 (1914): 372–75. In 1844, an article in *Medical News* explained "the organs of both sexes co-existing in the early stages of development, but one set alone in general developing afterwards at the expense of the antagonist set." See *Medical Examiner* 7 (1844): 47.

8. D. Berry Hart, "On Some Points in Regard to the Conditions of the Human Male and Female Usually Termed Hermaphrodism and Pseudohermaphrodism," *International Clinics* 4 (1915): 135–44, quotation on 135. Hart did not like the term pseudohermaphrodism, which usually described either a person with female external genitalia and male testes or vice versa. He preferred to call this "atypical sex-ensemble."

9. Alice Domurat Dreger, *Hermaphrodites and the Medical Invention of Sex* (Cambridge, Mass.: Harvard University Press, 1998), 158.

10. Alison Redick, "American History XY: The Medical Treatment of Intersex, 1916–1955" (PhD diss., New York University, 2004).

11. C. D. Creevy anticipated the significance of hormones to determining one's true sex as early as 1933: "It may be that in the future our knowledge of the sex hormones will advance sufficiently to permit us to rely upon them in determining the sex. Until that time, however, our only adequate method must remain biopsy of the gonads." See Creevy, "Pseudohermaphroditism," *International Surgical Digest* 16 (October 1933): 195–212, quotation on 209. Estrogen was isolated in 1929; progesterone in 1932. Some doctors promoted hormones as a key factor in the 1930s but did not abandon gonads as the primary indicator of true sex. See Raleigh R. Huggins, Mortimer Cohen, and Boyd Harden, "True Hermaphroditism in Man, with an Endocrinologic Study," *American Journal of Obstetrics and Gynecology* 34 (1937): 136–41; and Daniel R. Mishell, "Familial Intersexuality," *American Journal of Obstetrics and Gynecology* 35 (1938): 960–70. On the history of sex hormones, see Nelly Oudshoorn, *Beyond the Natural Body: An Archeology of Sex Hormones* (London: Routledge, 1994). On the development of hormones in relation to the management of menopause, see Judith A. Houck, *Hot and Bothered: Women, Medicine, and Menopause in Modern America* (Cambridge, Mass.: Harvard University Press, 2006). On the history of heredity as a factor in sex development, see Jane Maienshein, "What Determines Sex? A Study of Converging Approaches, 1880–1916," *ISIS* 75 (1984): 457–580.

12. Arthur Edmunds, "Pseudo-Hermaphroditism and Hypospadias," *Lancet* 1 (February 13, 1926): 323–27, quotation on 323. Though this paper was published in England, it exemplified the attention to social justifications for surgery found in America as well.

13. Ibid., 323. In their case, Drs. Robert Moehlig and Norman Allen described their patient who had been raised a girl but rechristened at age twelve after doctors

decided he was male. "He expressed a desire to marry in the future but felt embarrassed because he had to sit down to urinate. This was his only complaint" (1938). Robert C. Moehlig and Norman M. Allen, "Intersexuality," *Journal of the American Medical Association* 113 (1939): 1938–39.

14. In 1928, a man with an imperforate penis, more like a clitoris which lacked mobility, according to his doctor, had an operation to relieve him of painful erections. "The results of this procedure were most gratifying to the patient, as they converted him from a dejected, impotent individual into a happy and sexually active one" (111). A. P. Vastola, "Embryonal Carcinoma of Abdominal Testis in a Pseudohermaphrodite," *Journal of the American Medical Association* 101 (1933): 111–14.

15. Charles-Marie Debierre, translated by J. Henry C. Simes, *Malformations of the Genital Organs of Woman* (Philadelphia: P. Blakiston's Son, 1905), 118, 119. As it was presumed that coitus would only be justified within marriage, Debierre said that if marriage was impossible for one reason or another, then he believed surgeons should refuse the request because there would be no sound reason for it.

16. Until recently, doctors only occasionally paid attention to sexual feeling when considering these operations. Instead they focused on removing the clitoris (and sometimes enlarging the vagina) so that heterosexual penetration could occur. Penetration was more important to preserve than sexual pleasure. See, for example, William H. Rubovits and William Saphir, "Intersexuality," *Journal of the American Medical Association* 19 (1938): 1823–26. The doctors removed the woman's penis and testicles and enlarged her vagina, at her request. "Daily digital and instrumental dilations were partly successful in maintaining a sexually serviceable vagina" (1825).

17. J. Mark O'Farrell, "Hereditary Hermaphroditism: A Report of Three Cases," *Journal of the American Medical Association* 104 (1935): 1968–72, quotation on 1970.

18. Ibid., 1970.

19. Ibid., 1971.

20. Ibid.

21. Wendy Kline, *Building a Better Race: Gender, Sexuality, and Eugenics from the Turn of the Century to the Baby Boom* (Berkeley: University of California Press, 2001), 141–56.

22. William J. Robinson, *Woman: Her Sex and Her Love Life* (New York: Eugenics Publishing, 1917), 201, 216; Robinson, *Sexual Impotence: A Practical Treatise on the Causes, Symptoms, and Treatment of Sexual Impotence and Other Sexual Disorders in Men and Women* (New York: Critic and Guide, 1922); and Theodore H. Van de Velde, *Ideal Marriage: Its Physiology and Technique* (Lon-

don: William Heinemann, 1928). On the relationship between psychiatric illness and marriage, see Eugen Kahn, "Psychiatric Contraindications to Marriage," *Connecticut State Medical Journal* 6 (1941): 684–86.

23. Anon., "Hermaphrodism" *Medical Record* 41 (1892): 115.

24. Creevy, "Pseudohermaphroditism," 195–212, quotation on 196.

25. Milton Helpern, "True Hermaphroditism," *Archives of Pathology* 28 (1939): 768–69.

26. Ibid., 769.

27. Anon. [William J. Robinson?], "An Essay on Sexual Inversion, Homosexuality, Hermaphroditism" *Critic and Guide* 25A (July 1923): 247–60, quotation on 250.

28. Hugh Hampton Young, *Genital Abnormalities, Hermaphroditism and Related Adrenal Diseases* (Baltimore: Williams and Wilkins, 1937), 48. Often referred to as the father of modern American urology, Young practiced at Johns Hopkins from 1898 until he retired in 1942. See also Hugh H. Young and David M. Davis, *Young's Practice of Urology: Based on a Study of 12,500 Cases* (Philadelphia: W. B. Saunders, 1926).

29. Young, *Genital Abnormalities*, 51.

30. Alison Redick calls Young's approach "haphazard," and I don't disagree. Redick, *American XY*, 93. Some parents refused any surgery, including exploratory or postmortem, for their babies. In Samuel Levy's case, the parents would not agree to hospitalization, and they were convinced that the photographs taken of their child (the grandmother had pronounced the baby "half boy and half girl") hastened its death at ten days old (260). They would not permit an autopsy either. See Levy, "Pseudo-Hermaphrodism: Report of a Case in a New-Born," *Archives of Pediatrics* 47, no. 4 (April 1930): 259–62.

31. On the inadequacy of pediatric surgery in the early to mid-twentieth century, see C. Everett Koop, "Pediatric Surgery: The Long Road to Recognition," *Pediatrics* 92, no. 4 (October 1993): 618–21. The first modern textbook on the subject was published in 1941: William E. Ladd and Robert E. Gross, *Abdominal Surgery of Infancy and Childhood* (Philadelphia: W. B. Saunders, 1941).

32. For an interesting examination of how doctors make decisions and sometimes make mistakes, see Jerome Groopman, *How Doctors Think* (Boston: Houghton Mifflin, 2007).

33. In a discussion of adrenal tumors, the authors presented a 1903 case where a girl died shortly after surgery. The article reported several other cases with similar bad outcomes. They admitted that "even mild surgical procedures were sometimes fatal." Nonetheless, they concluded, "As a rule, children withstand surgical operations very well, and if the operation be done as soon as the condition is recognized the chance of success seems good enough to be used." See Henry D.

Jump, Henry Beates Jr., and W. Wayne Babcock, "Precocious Development of the External Genitals Due to Hypernephroma of the Adrenal Cortex," *American Journal of the Medical Sciences* 147, no. 4 (April 1914): 568–74, quotation on 574.

34. Francis R. Hagner and H. B. Kneale, "Pseudo Hermaphrodism or Complete Hypospadias," *Transactions of the American Association of Genito-Urinary Surgeons,* 15 (May 1922): 11–29, quotation on 12; William H. Haynes, "Pseudo Hermaphroditism," *Urological and Cutaneous Review* 38 (1934): 321–22.

35. Hugh Hampton Young, "Some Hermaphrodites I Have Met," *New England Journal of Medicine* 209 (1933): 370–75, quotation on 375. For another example of a doctor changing his mind about a child's "true sex," and the attendant shifts in hormonal and surgical treatments, see Harold O. Jones, "Pseudohermaphroditsm," *American Journal of Obstetrics and Gynecology* 35 (1938): 701–3.

36. Young, *Genital Abnormalities*, 84–91, esp. 84, 87, 91.

37. Ibid., 91.

38. It was not until the 1960s, with the rise in sex-change surgeries for transsexuals, that lawyers turned to establishing legal definitions for man and woman. To this day, states vary in their interpretation of legal gender status, privileging gonadal status, chromosomal status or external genitalia. See Meyerowitz, *How Sex Changed*, 245–53.

39. Nearly all of the published medical reports of this time included photographs of patients' gonadal tissue. Sometimes a picture of the patient's naked body, typically, but not always, with covered eyes, would be included as well.

40. William Quinby, "A Case of Pseudo-Hermaphrodism, with Remarks on Abnormal Function of the Endocrine Glands," *Bulletin of the Johns Hopkins Hospital* 27 (February 1916): 50–53. Young, *Genital Abnormalities* discusses Quinby's patient but says the boy was eleven years old in 1915 (347–62).

41. Quinby, "Case of Pseudo-Hermaphrodism," 52.

42. Young, *Genital Abnormalities and Hermphroditism,* 347–62.

43. Ibid., 353.

44. Ibid., 354.

45. J. B. Carnett, "A Case of Gynandrous Pseudo-Hermaphroditism," *Surgical Clinics of North America* 10 (1930): 1325–27, quotation on 1326.

46. Ibid.

47. James F. McCahey, "Hermaphroditism: A New Conception," *Archives of Pathology* 25 (1938), 927–35, quotation on 927–28.

48. James F. McCahey, "The Sexes and Hermaphroditism," *Journal of Urology* 42 (1939), 1130–34, quotation on 1133.

49. Bell, *Sex Complex,* 5.

50. Creevy, "Pseudohermaphroditism,"195–212, quotations on 211–12.

51. Emil Novak, "Sex Determination," 413–20, quotation on 413–14. See also Anon., "Should Orchidectomy Be Performed in Pseudohermaphroditism?" *Journal of the American Medical Association* 109 (1937), 884–85.

52. Louis Cohen, "A Case of Hermaphroditism," *Weekly Bulletin of the St. Louis Medical Society* 27 (1933): 509–11. Cohen explained the patient's sexual attraction to men vaguely, stating that it "can probably be explained on a mechanical basis and may also be the result of environment" (510). See also Anon., "Should Orchidectomy be Performed?" 884–85.

53. Novak, "Sex Determination," 414.

54. As recently as 2005, an independent group of intersex people, parents, and clinicians known as the Consortium on the Management of Disorders of Sex Development, organized by Alice Dreger, the ISNA (Intersex Society of North America) director of medical education, wrote clinical guidelines and a parents' handbook that promoted "patient-centered care," emphasizing open communication between intersex people, parents, and clinicians. Generally, the shift away from a paternalistic health-care model has led to disclosure of medical information to patients. Regarding intersex, see E. J. Sutton, J. Young, A. McInerney-Leo, C. A. Bondy, S. E. Gollust, and B. B. Bieseker, "Truth-Telling and Turner Syndrome: The Importance of Diagnostic Disclosure," *Journal of Pediatrics* 148, no. 1 (2006): 102–7.

55. See Consortium on the Management of Disorders of Sex Development, *Clinical Guidelines for the Management of Disorders of Sex Development in Childhood* and *Handbook for Parents* (Rohnert Park, Calif.: Intersex Society of North America, 2006), both available at www.dsdguidelines.org.

56. Creevy, "Pseudohermaphroditism," 211. See also Gerald K. Wooll, "Pseudo-Hermaphroditism; Case Report," *Wisconsin Medical Journal* 39 (1931): 454–56; the patient, a forty-nine-year-old woman, was discovered to have testicles. Wooll considered her male and lamented that "he" had been misdiagnosed early in life and thus was unable to enjoy "the possibility of a normal sexual life if some plastic work had been done at an early age." Wooll did not recommend that the person change sex as an adult; in fact, he wrote, "The patient was not informed of his sexual condition" (456). Similarly, Donald Kozoll did not tell his male patient that he had removed his uterus: "I had no desire to create doubt in a mind that probably could not fathom such developments" (583). Kozoll, "Pseudohermaphroditism: Report of Two Cases," *Archives of Surgery* 45 (1942): 578–95.

57. Novak, "Sex Determination," 415.

58. As quoted in J. P. Pratt, "Pseudohermaphrodism," *Transactions of the American Gynecological Society* 65, for the year 1940 (St. Louis: C. V. Mosby, 1941), 199–211, quotation on 211. Emphasis added.

59. Novak, "Sex Determination," 414.

60. Young, *Genital Abnormalities*, 71.

61. Young, "Some Hermaphrodites I Have Met," 373.

62. Young, *Genital Abnormalities*, 75.

63. G. Norman Adamson, "Hermaphroditism (Report of a Case)," *Clinical Medicine and Surgery* 49 (1933): 145–47, quotations on 145 and 147.

64. Ibid., 147.

65. Ibid.. See also Judson Gilbert, "Tumors in Pseudohermaphrodites: Review of Sixty Cases and a Case Report," *Journal of Urology* 48 (1942): 665–72.

66. Ralph C. Kell, Robert A. Matthews, and Albert A. Bockman, "True Hermaphroditism: Report of a Confirmed Case," *American Journal of the Medical Sciences* 197 (1939): 825–32, quotation on 827.

67. Ibid., 831.

68. Ibid., 827.

69. Ibid., 831. Emphasis added.

70. Young, "Some Hermaphrodites I Have Met," 370–75, quotations on 372–73. See also William H. Rubovits and William Saphir, "Intersexuality," *Journal of the American Medical Association* 10, no. 22 (1938): 1823–26. In this case, a thirty-nine-year-old unmarried woman with testicles, attracted to men, sought "relief from the annoying libido," by castration, clitoral removal, and vaginal enlargement. The doctors provided her with a "sexually serviceable vagina" at the same time that they reduced her sex drive, "which has made it possible for her to discontinue masturbation and resume her normal occupation" (1825).

71. James A. Betts, "A Hermaphrodite," *Atlantic Medical Journal* 29, no. 10 (July 1926): 686–87, quotation on 687.

72. Young, *Genital Abnormalities*, 142–48. For an analysis of the cultural representation of hermaphrodites, see Rachel Adams, *Sideshow U.S.A.* (Chicago: University of Chicago Press, 2001). The artwork that publicized hermaphrodites or "half and halfs" at carnivals did not conform to the reality of intersex bodies; in the pictures bodies were typically purported to be divided longitudinally, one side anatomically male, the other female. The accompanying text, however, did reflect contemporary medical interpretation, often emphasizing the "bisexual" theory, which allowed that each person contained male and female elements to varying degrees. For a comprehensive collection of online photographs and carnival advertisements of hermaphrodites, see www.sideshowworld.com/Blow-Off HH.html. See also Alice Domurat Dreger, "Jarring Bodies: Thoughts on the Display of Unusual Anatomies," *Perspectives in Biology and Medicine* 43:2 (Winter 2000): 161–72.

73. Walton Martin, "Pseudohermaphroditismus Masculinis," *Surgical Clinics of North America* 9 (1929): 535–44, quotation on 542.

74. Ibid., 540.

75. Ibid. See Catherine Kudlick, "Modernity's Miss-Fits: Blind Girls and Marriage in France and America, 1820–1920" in *Women on Their Own,* ed. Rudolph Bell and Virginia Yans (New Brunswick, N.J.: Rutgers University Press, 2008), 201–18.

76. Martin, "Pseudohermaphroditismus Masculinis," 543.

77. As quoted in J. P. Pratt, "Pseudohermaphrodism," 881.

78. On the ethics of secrecy, see A. Natarajan, "Medical Ethics and Truth Telling in the Case of Androgen Insensitivity," *Canadian Medical Association Journal* 154, no. 4 (1996): 568–70; and B. D. Kemp et al., "Sex, Lies and Androgen Insensitivity Syndrome" *Canadian Medical Association Journal* 154, no. 12 (1996): 1829–33; Alice Domurat Dreger, "'Ambiguous Sex'—or Ambivalent Medicine? Ethical Issues in the Treatment of Intersexuality," *Hastings Center Report* 28, no. 3 (1998): 24–36.

79. John Money, Joan Hampson, and John Hampson, "Hermaphroditism: Recommendations Concerning Assignment of Sex, Change of Sex, and Psychologic Management," *Bulletin of the Johns Hopkins Hospital* 97 (1955): 284–300, quotation on 290. Emphasis added.

Chapter Five: Psychology, John Money, and the Gender of Rearing in the 1940s, 1950s, and 1960s

Epigraph. Francis M. Ingersoll and Jacob E. Finesinger, "A Case of Male Pseudohermaphroditism: The Importance of Psychiatry in the Surgery of this Condition," *Surgical Clinics of North America* 47 (1947): 1218–25, quotation on 1224–25.

1. Doctors began testing urine samples to evaluate hormones in intersex cases in the 1930s. Raleigh Huggins, et al., believed that theirs was the first case, in 1937, in which both "male and female sex hormones have been simultaneously demonstrated in hermaphroditism." By 1940, doctors recognized that both androgens and estrogens were found in the human body for both sexes; Drs. Greenhill and Schmitz maintained, "Thus, the normal man or woman is to a slight degree an hermaphrodite." See Raleigh R. Huggins, Mortimer Cohen, and Boyd Harden, "True Hermaphroditism in Man, with an Endocrinological Study," *American Journal of Obstetrics and Gynecology* 34 (1937): 136–41, esp. 140. J. P. Greenhill and H. E. Schmitz, "Hermaphroditism and Sex Determination," *Western Journal of Surgery, Obstetrics, and Gynecology* 48 (1940): 36–41, quotation on 40.

2. Wm. James Barry, "A Case of Doubtful Sex," *Medical Examiner, and Record of Medical Science* 10 (May 1947): 308–9, quotation on 309.

3. See Heino Meyer-Bahlburg, "Gender Assignment in Intersexuality," *Journal of Psychology and Human Sexuality* 10, no. 2 (1998): 1–21.

4. Though Joan and John Hampson coauthored these articles, John Money elaborated upon them in countless later essays over the next three decades, and so his name has come to be primarily associated with the protocols they outlined in the 1950s. By 1958, Money's ideas were already incorporated into a textbook: Howard W. Jones and William Wallace Scott, *Hermaphroditism, Genital Anomalies and Related Endocrine Disorders* (Baltimore: Williams and Wilkins, 1958), esp. 49–50.

5. L. L. Langness, *The Study of Culture*, rev. ed. (Novato, Calif.: Chandler and Sharp, 1987), 99–137; Margaret Mead, *Coming of Age in Samoa* (New York: Morrow, 1928); Mead, *Sex and Temperament in Three Primitive Societies* (New York: Morrow, 1935); Sigmund Freud, *Totem and Taboo: Resemblances between the Psychic Lives of Savages and Neurotics* (1919; Harmondsworth, Middlesex: Penguin, 1938).

6. Jacob E. Finesinger, Joe V. Meigs, and Hirsh W. Sulkowitch, "Clinical, Psychiatric and Psychoanalytic Study of a Case of Male Pseudohermaphroditism," *American Journal of Obstetrics and Gynecology* 44 (1942): 310–17, quotation on 311.

7. In proving the existence of a "true hermaphrodite," some doctors included a mixture of social characteristics, such as a girl who was "excitable and nervous" and who liked to play with dolls, but who also showed an interest in mechanical things. See David M. Davis and Lewis C. Scheffey, "A Case of True Hermaphroditis," *Journal of Urology* 56 (1946): 715–17.

8. Finesinger, Meigs, and Sulkowitch, "Clinical, Psychiatric and Psychoanalytic Study," 317. Several reports mention the application of nail polish and other cosmetics as proof of femaleness. See, for example, Robert C. Moehlig, "Intersexuality Associated with Malignant Intra-Abdominal Teratoma of the Seminoma Type," *Journal of Clinical Endocrinology* 2 (1942): 257–61; and Seymour F. Wilhelm, "Resection of Hyperplastic Adrenal Glands for Female Pseudohermaphroditism," *Journal of the Mount Sinai Hospital, New York* 14 (1947): 679–87.

9. Finesinger, Meigs, and Sulkowitch, "Clinical, Psychiatric and Psychoanalytic Study," 316.

10. Ibid.

11. Ingersoll and Finesinger, "Case of Male Pseudohermaphroditism," 1218–25, quotation on 1220. Emphasis added. This view was widely shared. J. P. Pratt wrote in 1940, "It would seem that the judgment of the psychiatrist should be given a great deal of weight in any consideration of the treatment of pseudohermaphrodites." See Pratt, "Pseudohermaphroditism," 199–211, quotation on 209.

The typical hormonal assays tested urinary 17-ketosteroids and follicle-stimulating hormones.

12. Ingersoll and Finesinger, "Case of Male Pseudohermaphroditism," 1222.

13. Ibid.

14. See Barbara Welter, "The Cult of True Womanhood," *American Quarterly* 18 (Summer 1966): 151–74.

15. In fact, doctors William H. Perloff and Morris W. Brody explicitly denied this troubling possibility: "Shall one follow the wishes of the patient himself, or shall nature's ineffective intent be the motivating influence? Obviously, the final decision cannot be up to the patient." See Perloff and Brody, "Clinical Management of a Mal Pseudohermaphrodite," *Postgraduate Medicine* 10, no. 4 (October 1951), 337–40, quotation on 339. On the origination of the concept of sex hormones, see Nelly Oudshoorn, *Beyond the Natural Body: An Archeology of Sex Hormones* (London: Routledge, 1994).

16. Ingersoll and Finesinger, "Case of Male Pseudohermaphroditism," 1225.

17. Ibid.

18. Ibid., 1224.

19. G. Cotte, "Plastic Operations for Sexual Ambiguity," *Mt. Sinai Medical Journal* 14 (1947): 170–74. This is a reprint and translation of his article originally published in *Gynecologie et Obstetrique* in 1941.

20. Ibid., 174.

21. Ibid.

22. Albert Ellis, "The Sexual Psychology of Human Hermaphrodites," *Psychosomatic Medicine* 7 (1945): 108–25, quotation on 119.

23. Ibid., 108.

24. Ibid., 119.

25. Perloff and Brody, "Clinical Management," 337–40, quotation on 338.

26. Ibid., 339. For a similar discussion that speculates on the homosexuality of women with internal male organs who are attracted to men, see Emil Witschi and William F. Mengert, "Endocrine Studies on Human Hermaphrodites and Their Bearing on the Interpretation of Homosexuality," *Journal of Clinical Endocrinology* 2, no. 5 (May 1942): 279–86.

27. Perloff and Brody, "Clinical Management," 339.

28. H. S. Crossen, "A Problem in Sex Classification," *American Journal of Obstetrics and Gynecology* 38 (1939): 123–29.

29. Ibid., 129.

30. Greenhill and Schmitz, "Hermaphroditism and Sex Determination," paper presented at the annual meeting of the American Association of Obstetricians, Gynecologists, and Abdominal Surgeons in Hot Springs, Va., September 7–9, 1939.

31. Ibid., 36.

32. Ibid. Drs. Ira R. Sisk and Philip M. Cornwell detailed several cases in which the patients' psychology and anatomy contradicted their gonads: "It would be folly to attempt to change them to conform anatomically with the histology of their gonads" (736). Sisk and Cornwell, "Pseudohermaphroditism," *Journal of Urology* 47 (1942): 721–37.

33. Greenhill and Schmitz, "Hermaphroditism and Sex Determination," 41. Dr. George H. Bunch, editor of the *Southern Medical and Surgery* agreed: "To perform only such operations as will assist a person to continue to live as of the sex already established for the individual seems best" (720). Bunch, "Corrective Surgery in Hermaphroditism," *Southern Medicine and Surgery* 102 (1940): 720.

34. Daniel Chanis, "Some Aspects of Hermaphroditism," *Journal of Urology* 47 (1942): 508–14, quotation on 513. Judson B. Gilbert believed that most patients wanted to continue living in the gender they were used to, even after they found out about contradictory gonads. He hoped that in the future, endocrine studies, rather than anatomy, would provide more complete answers. See Judson B. Gilbert, "Tumors in Pseudohermaphrodites: Review of Sixty Cases and a Case Report," *Journal of Urology* 48 (1942): 665–72.

35. Louis E. Fazen, "Female Intersex: Report of an Unusual Case," *Wisconsin Medical Journal* 48 (1949): 1077–78 (emphasis added). For an example of a case in which doctors evaluated the patient's gonadal status in addition to her psychological state, see Minnie B. Goldberg and Alice F. Maxwell, "Male Pseudohermaphroditism: Proved by Surgical Exploration and Microscopic Examination," *Journal of Clinical Endocrinology* 8 (1948): 367–79; see also Wilhelm, "Resection of Hyperplastic Adrenal Glands for Female Pseudohermaphroditism."

36. Leo F. Bleyer, "Pseudohermaphroditism: Report of Two Cases in the Same Family," *American Journal of Surgery* 76 (1948): 448–52, quotation on 449.

37. Charles Hooks, "Clinical Aspects of Intersexuality," *Journal of Urology* 62 (1949): 529–34, quotation on 533.

38. Grace H. Dicks and A. T. Childers, "The Social Transformation of a Boy Who Had Lived His First Fourteen Years as a Girl: A Case History," *American Journal of Orthopsychiatry* 4 (1944): 507–17.

39. Ibid., 508. Blaming the mother for raising a child in the wrong gender appeared several times in the doctors' accounts from this period and might be interpreted as a modernized version of impugning maternal imagination, as discussed in chapter 1. It is also consistent with what the author Philip Wylie termed "momism," the idea that negative child-rearing practices of American mothers could turn their children into effeminate male homosexuals. Psychiatrist Edward Strecker gave scientific credence to the notion and included mothers' ability to raise man-hating lesbians as well. See Wylie, *Generation of Vipers* (New York:

Farrar and Rinehart, 1942); Edward A. Strecker, *Their Mother's Sons: The Psychiatrist Examines an American Problem* (New York: J. B. Lippincott, 1946); Edward A. Strecker and Vincent T. Lathbury, *Their Mother's Daughters* (Philadelphia: J. B. Lippincott, 1956).

40. Dicks and Childers, "Social Transformation of a Boy," 509.

41. Ibid., 512.

42. Ibid., 510.

43. Ibid., 512.

44. The question of whether or not social rearing is enough to ensure a matching gender identity has been considered by transsexuals as well. Transsexuals (those who seek hormones and sex reassignment surgery to change their gender presentation) want to change their gender, despite years of social conditioning in the gender they were assigned at birth. Having lived their lives as boys or girls, they look for ways to match their outward appearance with the gender they feel themselves to be. Though this is not universally articulated as such, most transpeople want to be recognized or read by others in the same way that they see themselves. Having been reared as a boy, for example, when one feels like a girl, often makes for a miserable childhood and unsatisfied adulthood, and does not at all instantiate a stable gender identity as male. James Green, a prominent transsexual activist, explained his transition from female to male this way, "If gender was socially constructed, I would have turned out to be a girl. People told me all my life I was a girl, and I tried. I really tried. You think this was easy? I could have saved a lot of money and a lot of heartache if I didn't have to do this. But it was the only way I could be who I was." See David Tuller, "A Self-Made Man," *San Francisco Chronicle*, September 21, 1997.

45. Dicks and Childers, "Social Transformation of a Boy," 514.

46. Catharine Cox Miles, "Psychological Study of a Young Adult Male Pseudo-hermaphrodite Reared as a Female," in *Studies in Personality, Contributed in Honor of Lewis M. Terman*, ed. Quinn McNemar and Maud Merrill (New York: McGraw-Hill, 1942): 209–27, quotation on 210.

47. Ibid., 211.

48. Ibid., 213.

49. Ibid., 215, 217.

50. On the significance of psychometricians, particularly Terman and Miles, see Jennifer Terry, *An American Obsession: Science, Medicine, and Homosexuality in Modern Society* (Chicago: University of Chicago Press, 1999), 168–77.

51. Lewis M. Terman and Catherine Cox Miles, *Sex and Personality: Studies in Masculinity and Femininity* (New York: McGraw Hill, 1936); E. Lowell Kelly, Catharine Cox Miles, and Lewis M. Terman, "Ability to Influence One's Score on a Typical Pencil-and-Paper Test of Personality," *Character and Personality* 4

(1936): 206–15; Wendy Kline, *Building a Better Race: Gender, Sexuality, and Eugenics from the Turn of the Century to the Baby Boom* (Berkeley: University of California Press, 2001), 134–41.

52. Miles, "Psychological Study," 223.

53. Ibid., 225.

54. Ibid. For another case of a successful switch from female to male, see William B. Stromme, "True Hermaphroditism," *American Journal of Obstetrics and Gynecology* 55 (1948): 1060–64.

55. Rita S. Finkler, "Social and Psychological Readjustment of a Pseudohermaphrodite under Endocrine Therapy," *Journal of Clinical Endocrinology* 8 (1948): 88–96, quotation on 88.

56. On the development of testosterone therapy, see John Hoberman, *Testosterone Dreams: Rejuvenation, Aphrodisia, Doping* (Berkeley: University of California Press, 2005).

57. Finkler, "Social and Psychological Readjustment," 90, 93.

58. Leona M. Bayer, "Pseudo-hermaphrodism: A Psychosomatic Case Study," *Psychosomatic Medicine* 9 (1947): 246–55, quotation on 255.

59. Ibid., 255. Emphasis in original.

60. Charles Morgan McKenna and Joseph H. Kiefer, "Two Cases of True Hermaphroditism," *Journal of Urology* 52 (1944): 464–69, quotation on 467.

61. Ibid., 468.

62. Ibid.

63. John Money, Joan G. Hampson, and John L. Hampson, "Imprinting and the Establishment of Gender Role," *AMA Archives of Neurology and Psychiatry* 77 (March 1957): 333–36, quotation on 334. Emphasis added.

64. John W. Money, "Hermaphroditism: An Inquiry into the Nature of a Human Paradox" (PhD diss., Harvard University), midyear, 1951–52, part 1, 53.

65. Ibid., part II, 82. Money's findings surprised some doctors but not others. In 1939, three doctors discussed a patient who had been raised as a boy for his four years, then as a girl until he was eighteen, and then decided to live as male again. "In view of the frequency of psychoses and neuroses attributed to sexual repressions and maladjustments in structurally normal individuals, it seems nothing short of miraculous that this patient, subjected constantly since early childhood to unusually severe psychic and emotional trauma, should be eminently sane and stable" (800). Parke G. Smith, James R. Mack, and Maynard Murray, "A Case of True Hermaphroditism," *Journal of Urology* 41 (1939) 780–800.

66. See, for example, the discussion of teasing with respect to girls growing up with "a large phallus," in Joan G. Hampson, "Hermaphroditic Genital Appear-

ance, Rearing and Eroticism in Hyperadrenocorticism," *Bulletin of the Johns Hopkins Hospital* 96 (1955): 265–73, quotation on 271.

67. Keith L. Moore, Margaret A. Graham, and Murray L. Barr, "The Detection of Chromosomal Sex in Hermaphrodites from a Skin Biopsy," *Surgery, Gynecology and Obstetrics* 96 (1953): 641–48, quotation on 641.

68. Hooks, "Clinical Aspects of Intersexuality," 528.

69. Suzanne Kessler interviewed several doctors who concurred that Money's theory "has taken on the character of gospel." One doctor conceded, "and I don't know how effective it really is." See Suzanne Kessler, *Lessons from the Intersexed* (New Brunswick, N.J.: Rutgers University Press, 1998), 15. Today, some doctors are admitting that they rather blindly accepted Money's protocols years ago and have more recently changed their views. See Jorge J. Daaboul, M.D., "Does the Study of History Affect Clinical Practice? Intersex as a Case Study: The Physician's View," www.isna.org/articles/daaboul_history.

70. In 1951, Dr. Frank Hinman still espoused relying on gonads but recognized that the specifics of certain conditions militated against treating all cases "according to their gonadal sex." In order to prevent "needless phallic sacrifice," he suggested that five aspects must be weighed in the decision: surgical, endocrinologic, psychologic, moral, and legal. See Hinman, "Advisability of Surgical Reversal of Sex in Female Pseudohermaphroditism," *Journal of the American Medical Association* 146 (1951); 423–29, quotation on 423.

71. Alison Redick makes the point in her dissertation that treatment decisions were actually made by Lawson Wilkons, director of the Endocrine Clinic and specialist in pediatric endocrinology. Money and the Hampsons provided psychological counseling. See Redick, "American XY: The Medical Treatment of Intersex, 1916–1955" (PhD diss., New York University, 2004), 238.

72. Joan G. Hampson, "Hermaphroditic Genital Appearance," 265. Hampson used "psychosexual orientation" and "gender role" synonymously to mean what we would call today "gender identity," one's sense of oneself as male or female. On Money's use of the terms "gender role" and his later switch to "gender identity/role," see John Money, *A First Person History of Pediatric Psychoendocrinology* (New York: Plenum, 2002), 36–37.

73. Joan G. Hampson, "Hermaphroditic Genital Appearance," 266.

74. Ibid., 265.

75. See Money, "Hermaphroditism," passim. For an analysis of the work of Money and the Hampsons as well as some of the intersex cases of the 1930s and 1940s, see Beatrice L. Hausman, *Changing Sex: Transsexualism, Technology, and the Idea of Gender* (Chapel Hill, N.C.: Duke University Press, 1995), 72–109.

76. Joan G. Hampson, "Hermaphroditic Genital Appearance," 267.

77. Ibid., 267.

78. Ibid., 271.

79. Ibid., 272.

80. John Money, Joan G. Hampson, and John L. Hampson, "Hermaphroditism: Recommendations Concerning Assignment of Sex, Change of Sex, and Psychologic Management," *Bulletin of the Johns Hopkins Hospital* 97 (1955): 284–300, quotation on 288.

81. Ibid., 288.

82. When Perloff and Brody were considering removing their patient's clitoris in 1951 at her request, they noted that the "center of sexual tension was in the 'clitoris' and not in the surgically created vagina," as they had hoped. They attributed this lack to the "interruption of nervous pathways incident to the surgery" but also, following Freudian theory, hinted that their patient was not truly female because she did not experience vaginal orgasms (340). See Perloff and Brody, "Clinical Management."

83. Money, Hampson, and Hampson, "Imprinting," 334.

84. Money, Hampson, and Hampson, "Hermaphroditism," 295. In a 1968 book, Money declared, "Despite the importance of the clitoris as a focus of erotic feelings, its removal does not abolish the capacity to reach orgasm and does not, so far as one can judge, reduce the feeling of sexual gratification. In the female as well as the male, it is remarkable how much erotic tissue can be removed without loss of erotic pleasure and the capacity to reach a sexual climax" (93). See Money, *Sex Errors of the Body: Dilemmas, Education, Counseling* (Baltimore: Johns Hopkins University Press, 1968).

85. Money, Hampson, and Hampson, "Imprinting," 334.

86. Money, Hampson, and Hampson, "Hermaphroditism," 295.

87. Ibid., 296.

88. Ibid., 288.

89. On the relationship between homosexuality, "perversion," and political subversion that characterized the period, see Terry, *An American Obsession,* 329–52.

90. Money, Hampson, and Hampson, "Hermaphroditism," 290, 292.

91. Money, Hampson, and Hampson, "Examination of Some Basic Sexual Concepts: The Evidence of Human Hermaphroditism," *Bulletin of the Johns Hopkins Hospital* 97 (1955): 309. In a later work, published in 2002, Money included notes from a 1955 conference that recalled his views on homosexuality and intersex. At that time he had believed, "The general statement we can make is that a child's psychologic outlook, sexually speaking—his psychosexual orientation—is a product of or is determined by the kind of life experiences that this child encounters and the kind of life experiences that he or she transacts. That

wording, you will notice, is careful not to talk about either heredity or environment, the one versus the other." See Money, *First Person History,* 21.

92. John Money, Joan G. Hampson, and John L. Hampson, "Sexual Incongruities and Psychopathology: The Evidence of Human Hermaphroditism," *Bulletin of the Johns Hopkins Hospital* 98 (1956): 56, 53.

93. Money, Hampson, and Hampson, "Hermaphroditism," 286.

94. Ibid., 290.

95. Money, *First Person History,* 24.

96. Money, Hampson, and Hampson, "Examination of Some Basic Sexual Concepts," 301–19, quotations on 307 and 308. The authors unconsciously echoed Dr. Little (see chapter 3) who criticized a female patient who refused surgery to make her more masculine as lacking "courage."

97. Money, Hampson, and Hampson, "Imprinting," 336.

98. Generally speaking, transsexuals do not have ambiguous genitalia or incongruity between genitals, hormones, chromosomes, and rearing. Historically, some transsexuals have tried to convince physicians that they were, in fact, intersexed, so that they could get the surgery and hormones that they wanted to effect gender reassignment. Today, there is a growing contingent of transsexuals who take the position that their intersex condition is invisible because something atypical in their brain necessitates the need for transsexual surgery. See Joanne Meyerowitz, *How Sex Changed: A History of Transsexuality in the United States* (Cambridge, Mass.: Harvard University Press, 2002); and Deborah Rudacille, *The Riddle of Gender: Science, Activism, and Transgender Rights* (New York: Pantheon Books, 2005).

99. Meyerowitz, *How Sex Changed,* 217.

100. Alice Domurat Dreger, *Intersex in the Age of Ethics* (Hagerstown, Md.: University Publishing Group, 1999), passim.

101. Money, Hampson, and Hampson, "Hermaphroditism," 291.

102. Joan G. Hampson, "Hermaphroditic Genital Appearance," 272. Hampson advocated a more open strategy for older patients. "One thing that needs the utmost emphasis, though, is that we have to consider in individuals who are old enough: What about the child's own preference? I think we doctors get rather into a bad habit of making decisions about what is the best thing to do for people, though with their consent admittedly."

103. Money, Hampson, and Hampson, "Hermaphroditism," 289.

104. See Dreger, *Intersex in the Age of Ethics.*

105. Sharon E. Preves, "For the Sake of the Children: Destigmatizing Intersexuality," in Dreger, *Intersex in the Age of Ethics,* 51–65, quotation on 55–56. See also Preves, *Intersex and Identity: The Contested Self* (Rutgers, N.J.: Rutgers University Press, 2003).

106. John Money, "Hermaphroditism and Pseudohermaphroditism," in *Textbook of Gynecologic Endocrinology*, ed. Jay J. Gold (New York: Hoeber, 1968), 461. See also Money, *Sex Errors of the Body: Dilemmas, Education, Counseling* (Baltimore: Johns Hopkins University Press: 1968), 61–62.

107. John Money, Tom Mazur, Charles Abrams, and Bernard F. Norman, "Micropenis, Family Mental Health, and Neonatal Management: A Report on 14 Patients Reared as Girls," *Journal of Preventive Psychiatry* 1 (1981): 17–27, quotation on 22.

108. Anita Natarajan, "Medical Ethics and Truth-Telling in the Case of Androgen Insensitivity Syndrome," *Canadian Medical Association Journal* 154 (1996): 568–70. For intersex adults' stories, see, for example, Cheryl Chase, "Hermaphrodites with Attitude: Mapping the Emergence of Intersex Political Activism," *GLQ: A Journal of Lesbian and Gay Studies* 4 (1998): 189–211; and Sherri Groveman, "The Hanukkah Bush: Ethical Implications in the Clinical Management of Intersex," in Dreger, *Intersex in the Age of Ethics*, 23–28.

109. Money, Hampson, and Hampson, "Hermaphroditism," 294.

110. John Money, Joan G. Hampson, and John L. Hampson, "The Syndrome of Gonadal Agenesis: Ovarian Agenesis and Male Chromosomal Pattern in Girls and Women: Psychological Studies," *Bulletin of the Johns Hopkins Hospital* 97 (1955): 207–26, quotation on 225.

111. Ibid., 224.

112. See John Colapinto, *As Nature Made Him: The Boy Who Was Raised as a Girl* (New York: Harper Collins, 2000). For a perceptive analysis of the case and its critiques, see Katrina A. Karkazis, *Fixing Sex: Intersex, Medical Authority, and Lived Experience* (Durham, N.C.: Duke University Press, 2008), esp. 69–77.

113. Simone de Beauvoir, *The Second Sex* (1949; New York: Random House, 1974), 301.

114. See, for example, John Money, *The Psychologic Study of Man* (Springfield, Ill.: Charles C. Thomas, 1957); Money, *Sex Errors of the Body*; John Money and Patricia Tucker, *Sexual Signatures: On Being a Man or a Woman* (Boston: Little, Brown, 1975).

115. Ultimately, David Reimer killed himself at age thirty-eight. On the complex reasons for his suicide, see Colapinto, "What Were the Real Reasons behind David Reimer's Suicide," *Slate*, June 3, 2004, slate.com/id/2101678/.

116. Milton Diamond and H. Keith Sigmundson, "Sex Reassignment at Birth: A Long Term Review and Clinical Implications," *Archives of Pediatrics and Adolescent Medicine* 151 (March 1997): 298–304. Available online at www.hawaii.edu/PCSS/online_artcls/intersex/mdfnl.html. See also Diamond's earlier criticism

of Money's theories: Diamond, "A Critical Evaluation of the Ontogeny of Human Sexual Behavior," *Quarterly Review of Biology* 40 (1965): 147–75. For Money's response to the Diamond and Sigmundson critique, see Money, *First Person History,* 71–76.

117. Marianne Shay, Letter to the Editor, *Time,* January 29, 1973. For a critique of Money as well as the relationship between scientific writing and cultural interpretation more generally, see Anne Fausto-Sterling, "How to Build a Man," in *Science and Homosexualities,* ed. Vernon Rosario (New York: Routledge, 1997), 219–55.

118. Milton Diamond, "A Critical Evaluation of the Ontogeny of Human Sexual Behavior," quotation on 150. On the scholarly arguments between Money, Diamond, and Zuger, see Anne Fausto-Sterling, *Sexing the Body: Gender Politics and the Construction of Sexuality* (New York: Basic Books, 2000), 66–71. Throughout the 1960s, there were a few dissenting voices regarding the efficacy of sex change in older patients. See, for example, M. Roth and J. R. B. Ball, "Psychiatric Aspects of Intersexuality," in *Intersexuality in Vertebrates Including Man,* ed. C. N. Armstrong and A. J. Marshall (London: Academic Press, 1964), 395–443. Roth and Ball argued that intersex people ought to be able to make their own decisions about their gender identities based on a variety of factors. See also Ian Berg, Harold H. Nixon, and Robert MacMahon, "Change of Assigned Sex at Puberty," *Lancet* 2 (1963): 1216–17; C. J. Dewhurst and R. R. Gordon, "Change of Sex," *Lancet* 2 (1963), 1213–16. For an extended discussion of activist opposition as well as challenges from the medical community, see Dreger, *Hermaphrodites,* 167–201; Karkazis, *Fixing Sex,* 237–77; Morgan Holmes, *Intersex: A Perilous Difference* (Selinsgrove, Penn.: Susquehanna University Press, 2008).

119. Bernard Zuger, "Gender Role Determination: A Critical Review of the Evidence from Hermaphroditism," *Psychosomatic Medicine* 32 (1970): 449–63, esp. 458–59.

120. Ibid., 463.

121. On the history of the intersex rights movement, see Karkazis, *Fixing Sex*; Holmes, *Intersex*; and Alice D. Dreger and April Herndon, "Progress and Politics in the Intersex Rights Movement: Feminist Theory in Action," *GLQ: A Journal of Lesbian and Gay Studies* 15, no. 2 (March 2009): 199–224. See also Suzanne Kessler's 1990 pioneering article, Kessler, "The Medical Construction of Gender: Case Management of Intersexed Infants," *Signs* 16 (1990): 3–26; and Fausto-Sterling, *Sexing the Body.*

122. Morgan Holmes, "Rethinking the Meaning and Management of Intersexuality," *Sexualities* 5, no. 2 (2002): 159–80, quotation on 174.

123. "Urologists: Agonize over Whether to Cut, Then Cut the Way I'm Telling

You," Oct. 14, 2004, Alice Dreger's Blog, *Intersex Society of North America,* www.isna.org. See also Dreger, *Intersex in the Age of Ethics.*

124. Fausto-Sterling, *Sexing the Body,* 67.

125. Money, et.al, "Micropenis, Mental Health, and Neonatal Management," 22.

126. Francis Wharton and Moreton Stillé, *Treatise on Medical Jurisprudence* (Philadelphia, 1855), 317.

Epilogue: Divergence or Disorder?

1. John P. Mettauer, M.D., "Practical Observations on those Malformations of the Male Urethra and Penis, termed Hypospadias and Epispadias, with an Anomalous Case," *American Journal of the Medical Sciences* 4 (July 1842): 43.

2. Richard Goldschmidt, "Intersexuality and the Endocrine Aspect of Sex," *Endocrinology* 1 (1917): 433–56.

3. Cheryl Chase, *Hermaphrodites Speak!* videocassette directed by Cheryl Chase (Rohnert Park, Calif.: Intersex Society of North America, 1996); Chase, "Hermaphrodites with Attitude: Mapping the Emergence of Intersex Political Activism," *GLQ: A Journal of Lesbian and Gay Studies* 4 (1998): 189–211.

4. Sharon Preves, "Out of the O.R. and Into the Streets: Exploring the Impact of Intersex Media Activism," *Research in Political Sociology* 13 (2004): 179–223.

5. See Arlene Baratz, Electronic Letters (September 11, 2006), adc.bmj.com/cgi/eletters/91/7/554#2590.

6. Alice D. Dreger and April Herndon, "Progress and Politics in the Intersex Rights Movement: Feminist Theory in Action" *GLQ: A Journal of Lesbian and Gay Studies* 15, no. 2 (March 2009): 199–224.

7. Alice Dreger, Cheryl Chase, Aron Sousa, Joel Frader, and Philip Grupposo, "Changing the Nomenclature/Taxonomy for Intersex: A Scientific and Clinical Rationale," *Journal of Pediatric Endocrinology and Metabolism* 18 (2005): 729–33.

8. Eric Vilain, "We Used to Call Them Hermaphrodites," *Genetic Medicine* 9 (2007): 65–66.

9. Cheryl Chase of ISNA and Barbara Thomas of the German group XY-Frauen were the two intersex adults who attended. Thomas's report of the conference can be found at www.aissg.org/PDFs/Barbara-Chicago-Rpt.pdf.

10. ISNA has closed its doors, but its founder, Cheryl Chase (now Bo Laurent) has opened Accord Alliance to continue its work, particularly in fostering collaboration between patients, families, and health-care teams.

11. Suzanne J. Kessler, *Lessons from the Intersexed* (New Brunswick, N.J.: Rutgers University Press, 1998), 32.

12. On the incidence of intersex, see the introduction, note 5.

13. Morgan Holmes, "Rethinking the Meaning and Management of Intersexuality," *Sexualities* 5 (2002): 159–80.

14. Rosemarie Garland Thomson, *Extraordinary Bodies: Figuring Physical Disability in American Culture and Literature* (New York: Columbia University Press, 1997).

15. Alice Domurat Dreger, *One of Us: Conjoined Twins and the Future of Normal* (Cambridge, Mass.: Harvard University Press, 2004).

16. Morgan Holmes, Rethinking," 159–80; Judith Butler, "Doing Justice to Someone: Sex Reassignment and Allegories of Transsexuality," *GLQ: A Journal of Lesbian and Gay Studies* 7 (2001): 621–36; Iain Morland, "Is Intersexuality Real?" *Textual Practice* 15 (2001): 527–47; M. Hird, "Gender's Nature: Intersexuals, Transsexuals and the 'Sex'/'Gender' Binary," *Feminist Theory* 1 (2000): 347–64; Alice Domurat Dreger, "Cultural History and Social Activism: Scholarship, Identities, and the Intersex Rights Movement," in *Locating Medical History: The Stories and Their Meaning*, ed. Frank Huisman and John Harley Warner, 390–409 (Baltimore: Johns Hopkins University Press, 2004).

17. See *Yellow for Hermaphrodite: Mani's Story*, 2004. DVD (Aukland, N.Z.: Greenstone Pictures); See generally the electronic letters in response to I. Hughes et al.'s article in *Archives of Disease in Childhood*, adc.bmj.com/cgi/eletters/91/7/554#2590.

18. Christine Matta, "Ambiguous Bodies and Deviant Sexualities: Hermaphrodites, Homosexuality, and Surgery in the United States, 1850–1904," *Perspectives in Biology and Medicine* 48 (2005): 74–83; Geertje Mak, " 'So We Must Go behind Even What the Microscope Can Reveal': The Hermaphrodite's 'Self' in Medical Discourse at the Start of the Twentieth Century," *GLQ: A Journal of Lesbian and Gay Studies* 11, no. 1 (2005): 65–94.

19. According to Meyer-Bahlburg, dexamethasone is thought to "reduce the degree of genital masculinization (with all its implications) in female newborns with classical CAH." The "implications" are spelled out in the article: body image problems, lack of maternal feeling, and lesbianism. See H. F. L. Meyer-Bahlburg, "What Causes Low Rates of Child-Bearing in Congenital Adrenal Hyperplasia," *Journal of Clinical and Endocrinological Metabolism* 84 (1999): 1844–47; Sharon Sytsma, "The Ethics of Using Dexamethasone to Prevent Virilization of Female Fetuses," in *Ethics and Intersex* ed. Sharon Sytsma (London: Springer, 2006); Monica J. Casper and Courtney Muse, "Genital Fixations," *American Sexuality Magazine*, http://nsrc.sfsu.edu/MagArticle.cfm?Article=595&PageID=0.

20. Alice Domurat Dreger, *Intersex in the Age of Ethics* (Hagerstown, Md.:

University Publishing Group, 1999); Iain Morland, "'The Glans Opens Like a Book': Right and Reading the Intersex Body," *Continuum: Journal of Media and Culture Studies* 19 (2005): 335–48.

21. Among the dissenters are some transsexual activists who think of themselves as intersexed. On the complicated relationship between transsexuality and intersex politics, see Dreger and Herndon, "Progress and Politics."

22. See Consortium on the Management of Disorders of Sex Development, *Clinical Guidelines for the Management of Disorders of Sex Development in Childhood* (Rohnert Park, Calif.: Intersex Society of North America, 2006); and *Handbook for Parents* (Rohnert Park, Calif.: Intersex Society of North America, 2006), both available at www.dsdguidelines.org.

23. Ibid.

24. David Cameron (August 2, 2006), and Milton Diamond and Hazel G. Beh (July 27, 2006), in Electronic Letters, http://adc.bmj.com/cgi/eletters/91/7/554#.

25. Consortium on Disorders of Sex Development, *Handbook for Parents* and *Clinical Guidelines*.

26. Patricia Donahue, David M. Powell, and Mary M. Lee, "Clinical Management of Intersex Abnormalities," *Current Problems in Surgery* 28 (1991), 513–79; see also Kessler, *Lessons From the Intersexed*; and Kessler, "The Medical Construction of Gender: Case Management of Intersexed Infants," *Signs* 16 (1990): 3–26.

27. Karkazis, *Fixing Sex: Intersex, Medical Authority, and Lived Experience* (Durham, N.C.: Duke University Press, 2008), esp. 265–90.

congenital adrenal hyperplasia, ix, 158, 163n4, 207n19
Conjugal Love; or, The Pleasures of the Marital Bed Considered in Several Lectures on Human Generation (Venette), 19
Consortium on the Management of Disorders of Sex Development, 158–159, 193n54
Cornwell, Philip M., 198n32
Cotte, G., 121–123
Cotton, John, 4
Creevy, Charles D., 89, 101, 103–104, 189n11
Cross, E., 71, 72
cross-dressing, 15–16, 31–32, 169n44, 174n14, 174n16
Crossen, H. S., 124–125
Culpeper, Nicholas, 3, 17

Davidge, John B., 168n31
Dean, Amos, 178nn49–50
Debierre, Charles-Marie, 86–87, 190n15
Denman, Thomas, 18
dexamethasone, 158, 207n19
Diamond, Milton, 150
Dicks, Grace H., 127–128
disability theory, 68, 73, 156, 157
disclosure. *See* secrecy/disclosure
disorders of sex development (DSD) (term), xiii, 153, 154, 155–159, 163n1
divergence of sex development (DSD) (term), xiii, 153–154, 159–160
divorce, 8–10, 167n23
Dreger, Alice, 51–52, 53, 85, 151, 158–159, 187n68, 193n54
DSD. *See* disorders of sex development; divergence of sex development
DuBois, Henry A., 68
duplicity, 24, 30–36; and class, 30, 32, 80; and cross-dressing, 174n16; and enlarged clitorises, 31, 174nn18–19; and existence of hermaphrodites, 32–33; and gender presentation, 30, 35; and gender roles, 32, 34, 36; and patient wishes, 80–81, 123; and psychological approaches, 123; and race, 80; and same-sex sexuality, 62, 69;

and sudden sex changes, 68; and "true sex" decisions, 31, 53, 60
Dyer, Mary, 3–5

early America, 1–22; and impotence, 8–10, 167n23; medical knowledge in, 2–3, 165n2; monstrosity in, 3–8; same-sex sexuality in, 15–22, 169n44, 172n75
early nineteenth century, 23–54; cross-dressing in, 174n13; duplicity in, 24, 30–36, 53, 174n19; existence of hermaphrodites in, 32–33, 41–44, 50–52, 177n47, 178nn49–50, 180n72; monstrosity in, 25–26, 173n5; normative heterosexuality in, 45–52, 179n61; professionalization of medicine in, 24, 28–29, 34; race, 36–40, 176n39; sympathy for hermaphrodites in, 24, 26–27, 28, 173n9; "true sex" decisions in, 48–49, 52–54, 179–180nn70–71
Edmunds, Arthur, 86
Ehrhardt, Anke A., 149
eighteenth century. *See* early America
Ellis, Albert, 123–124
Ellis, Havelock, 62
Emmet, Thomas Addis, 71
Enlightenment, 8
Essay on the Causes of the Variety of Complexion and Figure in the Human Species, An (Smith), 39
eugenics, 88
Eugenides, Jeffrey, 66, 164n4
European medical reports: and early American medical knowledge, 2–3; on gonadal definition of sex, 51–52, 53, 85; on impotence, 10; on mandatory choice of gender presentation, 11, 168n30, 179n70; on race, 20–21. *See also* Parsons, James
exhibitions, public, 69, 109, 110, 111, 185n40, 194n72

Farr, Samuel, 41, 175n28
Fausto-Sterling, Anne, ix, 151
Fazen, Louis E., 126
Female Marine, The, 35–36
feminism, 149, 157
Finesinger, Jacob E., 118–120, 121

Finkler, Rita, 133
Fissell, Mary, 7
5alpha-reductase deficiency, ix, 66,
    164n4
*Fixing Sex: Intersex, Medical Authority,
and Lived Experience* (Karkazis), 162
Forwood, W. S., 38–39
fraud. *See* duplicity
Freud, Sigmund, 117, 202n82

Gatti, John, 182n14
gender identity, 201n72. *See also* gender
    presentation; gender roles; "true sex"
    decisions
gender presentation: ambiguity in, 28;
    and duplicity, 30, 35; and existence of
    hermaphrodites, 175n28, 196n7; and
    gender roles, 60, 182n15; and gonadal
    definition of sex, 103; legal issues, 11,
    12–13; mandatory choice of, 2, 10–
    11, 13–14, 168n30, 169n35, 179n70;
    and normative heterosexuality, 46–47,
    49; and prostitution, 16; and same-
    sex sexuality, 65–66; and sexual
    desire, 64; and sudden sex changes,
    66–67; and "true sex" decisions, 52–
    53, 64, 76–77. *See also* gender roles;
    sex of rearing; "true sex" decisions
gender roles: and adult gender shifting,
    128, 129, 130, 131; and duplicity, 32,
    34, 36; and gender presentation, 60,
    182n15; malleability of, 135–139,
    149; and medical intervention, 70, 73;
    and psychological approaches, 116,
    118–119, 120, 196n8. *See also* gender
    presentation
Gilbert, Judson B., 198n34
Gilman, Sander L., 20–21
Goffe, J. Riddle, 78–80
Goldschmidt, Richard, 154
gonadal definition of sex, 83–85, 93,
    187n68; and "bisexuality" theory,
    84–85, 188–189nn5–7, 194n72; and
    castration, 102–103; and difficulty of
    "true sex" decisions, 91; doubts
    about, 98–102, 104, 111–112; Euro-
    pean medical reports on, 51–52, 53,
    85; and gender presentation, 103; and
    hormones, 189n11; and Money/

Hampson protocols, 135, 137, 139,
    201n70; and patient wishes, 96–97,
    107–109, 194n70; and psychological
    approaches, 121–123, 124–127,
    198nn32–33; and scientific advances,
    82, 96, 192n39
Green, James, 199n44
Greenhill, J. P., 125–126
Gross, Samuel D., 46–48, 70, 113, 152,
    179n61

Hagner, Francis R., 92
Hall, Thomas/Thomasine: and duplicity,
    30; and enlarged clitorises, 16,
    174nn18–19; and existence of her-
    maphrodites, 12–14; and mandatory
    choice of gender presentation, 10–11,
    13–14, 169n35; and same-sex sexual-
    ity, 16, 22; and "true sex" decisions,
    29
Hamilton, Alexander, 1, 2, 3, 17–18
Hamilton, William, 175n31
Hampson, Joan, 114, 137–138, 139,
    140, 146, 196n4, 201n72, 203n102.
    See also Money/Hampson protocols
Hampson, John, 114, 196n4. *See also*
    Money/Hampson protocols
Handy, William, 20, 171n65
Hardman, Belle, 67–68
Harlan, Richard, 177n47
Harris, S. B., 40
Hart, D. Berry, 189n8
hearsay. *See* anecdotal evidence
Helpern, Milton, 89–90
Henly, Mary, 15
hermaphrodites: mental, 57, 63, 65–66,
    183n22, 184n31, 184n33, 188n6,
    203n98; as neuters, 47, 59, 89, 181n4
hermaphrodites, existence of, 41–44,
    177n47, 188n73; and definitions, 1–2,
    57–58, 168n31, 180n72, 181n6; and
    duplicity, 32–33; and enlarged clito-
    rises, 17, 41; and Enlightenment, 8;
    and gender presentation, 175n28,
    196n7; and Hall case, 12–14; and
    intersex conditions, viii; and pseudo-
    hermaphrodites, 42–43, 59–60, 81,
    91, 99, 181n9; and sexual activity, 41,
    178n50; sympathy for, 8, 24, 26–27,

hermaphrodites (*cont.*)
28, 32, 61, 170n47, 173n9; and "true
sex" decisions, 43–44, 50–52, 58–59,
179n70, 181n10
hermaphrodite (term), vii–viii, xiii, 154–
155, 163n1
heterosexuality, normative, 45–52, 55,
104–105; and gender presentation,
46–47, 49; and medical intervention,
21, 22, 45–48, 49, 56, 68–69, 70, 71,
86–88, 179n61, 190n14, 190n16; and
Money/Hampson protocols, 141–143;
and "true sex" decisions, 48–49
Heustis, J. W., 40
Hinman, Frank, 201n70
Histoire des anomalies de l'organization
(Saint-Hilaire), 41–42
Hollick, Frederick, 39–40, 49, 50–51
Holmes, T., 48
homosexuality. *See* same-sex sexuality
Hooks, Charles, 137
hormones: and definitions of sex, 85,
116, 137, 189n11, 195n1; treatments
based on, 141
Howard, William Lee, 65–66
Hunter, Nan, 174n16
Hutchinson, Anne, 3, 4–5
hypospadias, ix; defined, 163n4; and
gonadal definition of sex, 96; and
impotence, 9; incidence of, 182n14;
and same-sex sexuality, 60; and sex of
rearing, 129; treatments for, 71, 92,
161, 172n74

identity, transgender, 65, 184n30,
184n33
immigration, 183n19
impotence, 8–10, 167n23
infertility. *See* sterility
Ingersoll, Francis M., 120, 121
intersex: congenital conditions, viii–ix,
163–164n4; incidence of, ix, 157,
164n5, 167n25, 182n14; as problem,
vii; term, xiii, 153, 154, 155, 158,
159, 163n1
intersex activists, xii; on gender roles,
73; on Money/Hampson protocols,
151–152; progress of, 162; on secrecy/
disclosure, 103, 145, 161, 193n54,

193n56; and terminology, 154–155,
157, 158, 163n1
Intersex Society of North America
(ISNA), 117, 151, 154, 156, 206n10,
258
Itinerarium (Hamilton), 1, 2, 3

Jacob, Giles, 18–20
Jewish law, 11, 13, 15, 168n29
John/Joan case, 148–150, 204n115
Jorgensen, Christine, 145
Juster, Susan, 15

Karkazis, Katrina, 162
Kessler, Suzanne, 137, 157, 201n69
Kiefer, Joseph H., 134, 135
King, Henry G., 28
Kline, Wendy, 88
Kneale, H. B., 92
Koop, C. Everett, 91

late nineteenth century, 55–81; existence
of hermaphrodites in, 57–59, 181n6,
181nn9–10; medical intervention in,
68–75; patient wishes in, 75–81;
same-sex sexuality in, 59–68; sexual
perversion in, 55
Laurent, Bo (Cheryl Chase), 73, 154,
206nn9–10
legal issues: cross-dressing, 15–16, 31–
32, 169n44, 174n14, 174n16;
divorce, 8–10, 167n23; gender presentation, 11, 12–13; immigration,
183n19; marriage, 11; same-sex sexuality, 16, 183n19; "true sex" decisions, 192n38; voting, 34
lesbianism. *See* same-sex sexuality
Lessons from the Intersexed (Kessler),
137
Levy, Samuel, 191n30
Lewis, Bransford, 183n23
Lewis, Deborah Francis, 14,
169nn38–39
Little, James, 75–78, 203n96
Long, J. W., 68
Luchins, Abraham, 165n2
Lydston, G. Frank, 59, 60, 62–63,
182n14, 183n22

Reimer, David, 148–150, 204n115
Robinson, William, 88
Ross, Robert A., 113
Roth, M., 205n118

Saint-Hilaire, Geoffroy, 41–42, 178n49
same-sex sexuality, 15–22, 59–68,
172n75; and adult gender shifting,
133; and anecdotal evidence, 61–62;
and castration, 69–70, 102–103,
193n52; and child-rearing, 198n39;
and dexamethasone treatment, 158,
207n19; and duplicity, 62, 69; and
enlarged clitorises, 15, 16, 18–21, 60–
61, 171n65; and gender presentation,
65–66; and laws against cross-dressing, 15–16, 169n44; and legal issues,
16, 183n19; and medical intervention,
55–56, 68–69, 70, 157–158; and
mental hermaphroditism, 57, 63, 65–
66, 183n22, 184n31; modern views
of, 56, 57; Money/Hampson protocols on, 142, 202–203n91; and moral
condemnation, 61, 62, 63; and patient
wishes, 105–107, 124; and pseudohermaphrodites, 59–60; and psychological approaches, 123–124; and
sexual pleasure, 60–61; and sudden
sex changes, 66–68; and "true sex"
decisions, 21, 73, 77–79, 105–107
Schmitz, H. E., 125–126
Schober, Justine, 164n7
Second Sex, The (Beauvoir), 149
secrecy/disclosure, 103–104, 160–161;
intersex activists on, 103, 145, 161,
193n54, 193n56; Money/Hampson
protocols on, 145–148, 203n102; and
terminology, 155
Sex and Personality: Studies in Masculinity and Femininity (Terman and
Miles), 131
sex changes, sudden, 19, 36, 63,
171n60; and gender presentation, 66–
7; and race, 36–39, 176n39; and
same-sex sexuality, 66–68
sex-change surgery. See transsexuals
sex chromosome mosaicism, ix, 164n4
sex development, disorders of (DSD)
(term), xiii, 153, 154, 155–159, 163n1

sex development, divergence of (DSD)
(term), xiii, 153–154, 159–160
sex of rearing: and adult gender shifting,
127, 128–129, 134–135; and maternal imagination, 198n39; and medical
intervention, 136–137; Money/
Hampson protocols on, 136–138,
139–145; and optimum gender of
rearing model, 116; and psychological
approaches, 119; and transsexuals,
199n44. See also gender presentation;
gender roles
sex reassignment surgery (SRS). See
transsexuals
sexual activity: and existence of hermaphrodites, 41, 178n50; and marriage, 190n15; and monstrosity, 7;
and moral condemnation, 84. See also
same-sex sexuality
sexual desire: and medical intervention,
46–47, 48; and normative heterosexuality, 47; and patient wishes, 80, 98;
and "true sex" decisions, 47–48, 53,
64, 180n77
sexual feeling, 141, 151, 190n16,
202n82, 202n84. See also clitorises:
excision of
Sex Variant Study, 133
Sharp, Jane, 3, 5–7, 18, 21, 171n56
Sigmundson, H. Keith, 150
Sisk, Ira R., 198n32
Smellie, William, 18
Smith, Samuel Stanhope, 39
Solomon, Leon L., 83
spurious hermaphrodites. See
pseudohermaphrodites
SRS (sex reassignment surgery). See
transsexuals
standing to urinate, 73, 75, 76, 86, 93,
111, 115, 129, 190n13
stereotypical gender roles. See gender
presentation; gender roles
sterility, 59, 123, 147, 148
Stewart, Gilbert H., 181n6
Stillé, Moreton, 40, 43–44, 48, 152
Strecker, Edward, 198n39
Studies in the Psychology of Sex (Ellis),
62
suicide, 98, 204n115